Mastering Angular Test-Driven Development

Build high-quality Angular apps with step-by-step instructions
and practical examples

Ezéchiel Amen AGBLA

Mastering Angular Test-Driven Development

Group Product Manager: Kaustubh Manglurkar

Publishing Product Manager: Urvi Shah

Book Product Manager: Aishwarya Mohan

Senior Editor: Rakhi Patel

Technical Editor: Reenish Kulshrestha

Copy Editor: Safis Editing

Indexer: Rekha Nair

Production Designer: Vijay Kamble

DevRel Marketing Coordinators: Anamika Singh and Nivedita Pandey

First published: August 2024

Production reference: 1090724

Published by Packt Publishing Ltd.

Grosvenor House

11 St Paul's Square

Birmingham

B3 1RB, UK

ISBN 978-1-80512-608-9

www.packtpub.com

To my mother, Lamatou, my father, Daniel, my brothers, Salomon and Schadrac, and my sister, Christelle, for their sacrifices and for demonstrating the power of determination.

To my sister-in-law, Anne and my nephew, Samuel, for their unfailing support.

To Irma, my godmother, Bernadin, my godfather, and my brother, Jean-David, for their invaluable advice and presence during difficult times.

Finally, to my wife, Béatrice, for the unconditional love she gives me, for supporting me during this journey, and for always being there when I need her.

– Ezéchiel Amen AGBLA

Contributors

About the author

Ezéchiel Amen AGBLA is passionate about web and mobile development, with expertise in Angular and Ionic. A certified Ionic and Angular expert developer, he shares his knowledge through courses, articles, webinars, and training.

He uses his free time to help beginners and spread knowledge by writing articles, including via Openclassrooms, the leading online university in the French-speaking world. He also helps fellow developers on a part-time basis with code reviews, technical choices, and bug fixes.

His dedication to continuous learning and experience sharing contributes to his growth as a developer and mentor.

I'd like to thank all the people who have been close to me and supported me, especially my wife, Béatrice, my brothers, my sister, and my parents.

I'd also like to thank the team at Packt for their help and support throughout the process.

About the reviewers

Afolabi Opakunle is a seasoned software engineer with extensive experience in Angular since version 6, specializing in designing, planning, and developing frontend web solutions tailored to product and design requirements. Leveraging Angular's modular framework, he has successfully led frontend development teams and contributed to enterprise projects for over four years, enhancing portfolio management processes.

He holds a certificate in Mobile Web Development from Google, where he gained a deep understanding of the Angular framework. His academic background includes a bachelor's degree in city/urban, community, and regional planning from Obafemi Awolowo University, Ile-Ife, Nigeria. With over five years in the software industry, he has focused on developing financial technology and business management solutions. Collaborating closely with project managers from different parts of the world, he has successfully developed and updated modules and features for various enterprise applications.

Passionate about solving business challenges through innovative technologies, he continually explores new solutions to enhance project outcomes. In his leisure time, he enjoys discovering new restaurants in Lagos with friends and team members and indulging in the vibrant culinary scene.

Dhritiman Tamuli Saikia is an experienced software developer as well as a trainer. He mostly works on Angular, ReactJS, Node.js, and Express.js. He has also trained numerous students on MEAN and MERN stack development in the last few years.

Sanjay Khadka, hailing from Nepal, is an IT engineer with over 15 years of experience in web/mobile application development and a passion for open source technology.

He has held roles as a developer, principal software engineer, team lead, project manager, and software and solution architect. Additionally, he also has experience teaching at training institutes.

He had also been a visiting project supervisor for university students and a trainer at Nepal Electricity Authority.

His technical skills include iOS, web and mobile development, SQL, MongoDB, PHP, Node.js, JavaScript, Angular, Rust, system design, DevOps, and Agile.

He also holds certifications such as Registered Scrum Master™, Registered Product Owner™, and AWS Certified Solution Architect.

I extend my sincere gratitude to the author for crafting this wonderful book. I am also thankful to Shagun Saini and Vaibhav Ahire for their invaluable assistance throughout the process. This comprehensive guide to test-driven development (TDD) is highly commendable and I strongly recommend it to anyone interested in TDD. I am looking forward to more books from the author in the future.

Table of Contents

5

Testing Angular Pipes, Forms, and Reactive Programming   81

Part 3: End-to-End Testing

6

Exploring End-to-End Testing with Protractor, Cypress, and Playwright   109

Part 4: Continuous Integration and Continuous Deployment for Angular Applications

10

Best Practices and Patterns for Angular TDD   197

11

Refactoring and Improving Angular Code through TDD   203

Preface

Hello, everyone! **Test-Driven Development** (**TDD**) is a software development process widely used in Angular development to ensure code quality and reduce the time spent debugging. TDD is an agile approach to software development that emphasizes the iterative writing of tests before writing the actual code. The TDD process comprises three steps:

- Red: In this initial phase, developers write a test for the functionality they aim to implement. Since there is no corresponding code yet, this test will initially fail, hence the term "Red" to indicate the failing state of the test.

- Green: Following the Red phase, developers write the minimum amount of code necessary to make the test pass. This phase aims to quickly move from a failing test (Red) to a passing test (Green), focusing on satisfying the test's conditions without necessarily optimizing the code.

- Refactor: After successfully passing the test, the code is then improved and optimized. This phase involves refining the code's design, structure, and efficiency while ensuring that the test remains green, i.e., continues to pass. Refactoring is crucial for enhancing code quality, maintainability, and performance without altering the external behavior of the code, as confirmed by the passing tests.

These steps are cyclically repeated for each new functionality, ensuring that every part of the software is thoroughly tested and backed by tests, thereby promoting high code quality, reducing bugs, and facilitating confident refactoring.

This book is a comprehensive guide that provides developers with the essential resources to improve their skills and deliver high-quality Angular applications. With a hands-on approach and real-world examples, it covers TDD concepts, techniques, and tools extensively, going beyond unit testing to explore Agular's pipe, form, and reactive programming testing. In this book, you'll learn how to validate and manipulate data using pipes, test Angular forms for input validation and user interaction, and handle asynchronous operations with reactive programming. In addition, you'll explore end-to-end testing using the Protractor, Cypress, and Playwright frameworks, gaining valuable insights into writing robust tests for web applications, covering navigation, element interaction, and behavior validation. You'll also explore the integration of TDD with CI/CD, learning best practices for automating tests, deploying Angular applications, and achieving faster feedback loops. With the help of concrete examples, best practices, and clear explanations, you'll be able to successfully implement TDD in your Angular projects by the end of this book.

Who this book is for

Mastering Angular Test-Driven Development is a comprehensive guide that provides developers with essential resources to enhance their skills and deliver high-quality Angular applications. With a practical approach and real-world examples, it extensively covers TDD concepts, techniques, and tools, going beyond unit testing to explore testing Angular pipes, forms, and reactive programming. In this book, you'll learn how to validate and manipulate data using pipes, test Angular forms for input validation and user interactions, and handle asynchronous operations with reactive programming. Additionally, you'll explore end-to-end testing using the Protractor, Cypress, and Playwright frameworks, gaining valuable insights into writing robust tests for web applications, covering navigation, element interaction, and behavior validation. You'll also explore integrating TDD with CI/CD, learning best practices for automating tests, deploying Angular applications, and achieving faster feedback loops.

With the help of practical examples, best practices, and clear explanations, you'll successfully implement TDD in your Angular projects by the end of this book.

What this book covers

Chapter 1, *Taking Your First Steps with TDD,*, has the objective of understanding and setting up test-driven development. We will analyze the concept through the problems it solves, the advantages it brings, the logic behind its implementation, and how it is relevant to the success of an Angular project.

Chapter 2, *Using Jasmine and Karma to Test Angular Applications*, has the goal of getting familiar with Jasmine and Karma and writing your first unit tests. You will learn more about the configurations to implement Jasmine and Karma in an Angular project.

Chapter 3, *Writing Effective Unit Tests for Angular Components, Services, and Directives*, looks at refactoring tests written for Angular components and writing tests for Angular services and directives used in the business logic of our project by respecting the TDD principles. The tests will be progressively refactored to highlight the gradual character of TDD.

Chapter 4, *Mocking and Stubbing Dependencies in Angular Tests*, focuses on the creation of mocks, which is an unavoidable aspect when testing Angular services and directives. At the same time, we will take the opportunity to complete and refactor the previous tests for our services and directives.

Chapter 5, *Testing Angular Pipes, Forms, and Reactive Programming*, explores how to test the pipes, reactive forms, and RxJS operators of an Angular project based on the principles of test-driven development.

Chapter 6, *Exploring End-to-End Testing with Protractor, Cypress, and Playwright*, looks at end-to-end testing – an approach to evaluate the functioning of a product in an end-to-end process. It is therefore wise to refer to it when you want to be sure of the quality of the application that is built. The objective is to understand it and learn more about the benefits of implementing it in an Angular project.

Chapter 7, Understanding Cypress and its Role in End-to-End Tests for Web Applications, looks at how the best tool for end-to-end testing in Angular is Cypress. We will discover together how to install it, configure it, and use it to perform our various component-oriented end-to-end tests.

Chapter 8, Writing Effective End-to-End Component Tests with Cypress, deals in more depth with the writing of end-to-end tests with Cypress – always taking a TDD approach because we will improve and refactor previous tests related to components while respecting best practices with Cypress.

Chapter 9, Understanding Continuous Integration and Continuous Deployment (CI/CD), discusses the topics of continuous integration and continuous development. The objective is to see how to set up a pipeline that will take care of verifying that software has been compiled and that all tests pass through automated jobs before being deployed on our remote server using GitLab CI/CD.

Chapter 10, Best Practices and Patterns for Angular TDD, focuses on best practices when implementing test-driven development in an Angular project and on the patterns that exist. The objective is to explore ways to converge toward clean code that's easy to maintain and less prone to bugs.

Chapter 11, Refactoring and Improving Angular Code through TDD, focuses on leveraging test-driven development (TDD) techniques to refactor and improve existing Angular code. By writing tests before making changes, you can ensure code quality and maintainability and confidently enhance functionality. TDD enables iterative cycles of writing tests, refactoring, and improving code based on test feedback, leading to more robust and efficient Angular applications.

To get the most out of this book

Dear readers, before you start reading the book, you should have a basic understanding of how to create applications using the Angular framework and the associated TypeScript language. Finally, you'll also need to know how to use Git and GitHub.

Software/hardware covered in the book	Operating system requirements
Angular 17 or later LTS	Windows, macOS, or Linux
TypeScript 5.1	Windows, macOS, or Linux
Node.js 18 or later LTS	Windows, macOS, or Linux

If you are using the digital version of this book, we advise you to type the code yourself or access the code from the book's GitHub repository (a link is available in the next section). Doing so will help you avoid any potential errors related to the copying and pasting of code.

Download the example code files

You can download the example code files for this book from GitHub at https://github.com/PacktPublishing/Mastering-Angular-Test-Driven-Development. If there's an update to the code, it will be updated in the GitHub repository.

We also have other code bundles from our rich catalog of books and videos available at https://github.com/PacktPublishing/. Check them out!

Conventions used

There are a number of text conventions used throughout this book.

Code in text: Indicates code words in text, folder names, filenames, file extensions, pathnames, dummy URLs, and user input.

Here is an example: "In relation to the contents of percent.pipe.spec.ts, this is what we have:"

A block of code is set as follows:

```
import { PercentPipe } from './percent.pipe';
describe('PercentPipe', () => {
  it('create an instance', () => {
    const pipe = new PercentPipe();
    expect(pipe).toBeTruthy();
  });
});
```

Any command-line input or output is written as follows:

```
$ ng g pipe percent -skip-import
$ ng test
```

Bold: Indicates a new term, an important word, or words that you see onscreen. For instance, words in menus or dialog boxes appear in **bold**. Here is an example: " Now we can save the file by clicking on the **Commit changes** button."

> **Tips or important notes**
> Appear like this.

Get in touch

Feedback from our readers is always welcome.

General feedback: If you have questions about any aspect of this book, email us at `customercare@packtpub.com` and mention the book title in the subject of your message.

Errata: Although we have taken every care to ensure the accuracy of our content, mistakes do happen. If you have found a mistake in this book, we would be grateful if you would report this to us. Please visit `www.packtpub.com/support/errata` and fill in the form.

Piracy: If you come across any illegal copies of our works in any form on the internet, we would be grateful if you would provide us with the location address or website name. Please contact us at `copyright@packt.com` with a link to the material.

If you are interested in becoming an author: If there is a topic that you have expertise in and you are interested in either writing or contributing to a book, please visit `authors.packtpub.com`.

Share Your Thoughts

Once you've read Mastering Angular Test-Driven Development, we'd love to hear your thoughts! Scan the QR code below to go straight to the Amazon review page for this book and share your feedback.

`https://packt.link/r/1805126083`

Your review is important to us and the tech community and will help us make sure we're delivering excellent quality content.

Download a free PDF copy of this book

Thanks for purchasing this book!

Do you like to read on the go but are unable to carry your print books everywhere?

Is your eBook purchase not compatible with the device of your choice?

Don't worry, now with every Packt book you get a DRM-free PDF version of that book at no cost.

Read anywhere, any place, on any device. Search, copy, and paste code from your favorite technical books directly into your application.

The perks don't stop there, you can get exclusive access to discounts, newsletters, and great free content in your inbox daily

Follow these simple steps to get the benefits:

1. Scan the QR code or visit the link below

https://packt.link/free-ebook/9781805126089

2. Submit your proof of purchase
3. That's it! We'll send your free PDF and other benefits to your email directly

Part 2: Writing Effective Unit Tests

3

Writing Effective Unit Tests for Angular Components, Services, and Directives   37

4

Mocking and Stubbing Dependencies in Angular Tests   59

Part 1:
Getting Started with
Test-Driven Development
in Angular

In this part, you'll get an overview of the benefits of **test-driven development** (TDD), the testing environment in an Angular project, writing unit tests with Jasmine, and configuring Karma.

This part has the following chapters:

- *Chapter 1, Taking Your First Steps with TDD*
- *Chapter 2, Using Jasmine and Karma to Test Angular Applications*

1

Taking Your First Steps with TDD

Test-driven development (TDD) is a software development process that's widely used in Angular development to ensure code quality and reduce time spent debugging. By writing automated tests before writing production code, developers can ensure that their code meets the desired specifications and can be easily modified and maintained over time.

In this chapter, we'll explore the early stages of TDD in Angular. We'll start by discussing the role of TDD in Angular development and how it can help improve the quality of your code. Then, we'll set up the development environment, which involves installing the necessary tools and dependencies.

Next, we'll create a new Angular project and explore the various files related to writing tests, including the spec files, which contain the actual tests, and the `karma.conf.js` file, which is used to configure the testing framework.

Throughout this chapter, I will provide examples and best practices for writing effective tests in Angular, such as using descriptive test names, testing all code paths, and using dummy data to simulate different scenarios.

By the end of this chapter, you'll have a solid understanding of the basics of TDD in Angular and how to start using it. Whether you're new to TDD or looking to improve your skills, this chapter will provide you with the knowledge and tools you'll need to create high-quality Angular applications using this approach.

We will cover the following topics:

- Understanding TDD and its role in Angular
- Setting up the development environment
- Creating a new Angular project
- Exploring different files related to writing tests

Technical requirements

To follow along with the examples and exercises in this chapter, you will need to have a basic understanding of Angular and TypeScript, as well as a code editor, such as Visual Studio Code, installed on your computer

All the code files for this chapter can be found at `https://github.com/PacktPublishing/Mastering-Angular-Test-Driven-Development/tree/main/Chapter%201`.

Understanding TDD and its role in Angular

In this section, we will explore the fundamentals of TDD and its role in Angular development. We'll start by discussing the benefits of using TDD in general, such as improved code quality, faster development cycles, and reduced debugging time. Then, we'll look at how TDD fits into the overall Angular development process and how it can be used to create scalable and maintainable applications.

What is Angular and TDD?

Angular is a popular open source JavaScript framework that's used for building complex web applications. It was developed by Google and is widely used by developers around the world. Angular provides a set of tools and features that make it easy to build dynamic, responsive, and scalable web applications.

Angular is a component-based framework that allows developers to build reusable UI components. These components can be combined to create complex user interfaces, making it easy to maintain and extend the application. Angular also provides built-in support for testing, making it easy to write and execute tests for Angular applications.

TDD is a dynamic methodology in software development that prioritizes incrementally creating tests before the code is implemented. The TDD process revolves around a structured sequence known as the **red-green-refactor** cycle, which consists of the following stages:

- **Writing a failing test**: Initiate the cycle by crafting a test that intentionally fails. This test serves as a specification for the desired functionality.

- **Prohibiting overly complex tests**: Emphasize the creation of tests that are only as intricate as necessary. Avoiding unnecessary complexity ensures that tests remain focused on specific functionalities, enhancing clarity and maintainability.

- **Minimizing code implementation**: Write the minimum code required to pass the failing test. This minimalist approach ensures that code is dedicated to fulfilling the specified requirements.

The iterative nature of TDD unfolds as follows: writing a failing test, implementing the code to pass the test, and refactoring the code to enhance code design and maintainability. This iterative loop persists until the entire code base is complete.

TDD's unique approach to writing tests before code execution serves a dual purpose. First, it guarantees code correctness by aligning with predefined test requirements. Second, it fosters the creation of clean, maintainable, and adaptable code. Developers are encouraged to adhere to best practices, resulting in code that is easily comprehensible, modifiable, and extensible throughout the software development life cycle.

Now that we know what Angular is and looked at the benefits of using the TDD approach, let's understand the red-green-refactor cycle.

The red-green-refactor cycle

The red-green-refactor cycle is a fundamental concept in TDD. It serves as a robust and systematic methodology in software development that offers developers a structured framework for incremental progress. This approach is designed to break down the development process into discrete, manageable steps, guaranteeing code correctness and alignment with predefined test requirements. Now, let's delve into the technical nuances of each phase – red, green, and refactor – within this iterative process:

- **Red – writing a failing test**:

    The first step in the red-green-refactor cycle is to write a failing test. The test should define the desired behavior of the code and should be written in a way that it fails initially. This is called the "red" step because the test is expected to fail.

- **Green – writing code to pass the test**:

    The second step is to write the code that will make the test pass. The code should be minimal, and it should only be written to make the test pass. This is called the "green" step because the test is expected to pass.

- **Refactor – improving code without changing functionality**:

    Once the test has passed, the developer can refactor the code to enhance its design, readability, and maintainability by eliminating duplication, simplifying the code, and improving its readability. The key is to make improvements without altering the functionality covered by the test.

In the next section, we will take a look at the benefits of the red-green-refactor cycle.

Benefits of the red-green-refactor cycle

The red-green-refactor cycle has several benefits, including the following:

- **Enhanced code quality**: The red-green-refactor cycle ensures that the code is correct, reliable, and meets the requirements predefined in the tests
- **Accelerated development**: The red-green-refactor cycle allows developers to catch errors early in the development process, which saves time and reduces the cost of fixing bugs
- **Better collaboration**: The red-green-refactor cycle encourages collaboration between developers, testers, and other stakeholders, which improves communication and helps to ensure that everyone is on the same page
- **Simplified maintenance**: The red-green-refactor cycle produces code that is easier to maintain and extend, which reduces the cost and effort of future development

By using the red-green-refactor cycle, developers can build reliable, maintainable, and scalable software applications.

Next, we'll learn how TDD is an important asset when it's in use.

The role of TDD in Angular development

TDD plays a critical role in Angular development. By writing tests first, developers can ensure that the code is correct and meets the requirements defined in the tests. This ensures that the application is reliable and reduces the risk of bugs and errors. TDD also encourages developers to write clean, maintainable code that is easy to modify and extend, making it easier to maintain and update the application over time.

Angular provides built-in support for testing, making it easy to write and execute tests for Angular applications. The Angular testing framework provides a set of tools and features that make it easy to write unit tests, integration tests, and end-to-end tests for Angular applications. These tests can be run automatically as part of the build process, ensuring that the application is always tested and reliable.

In the next section, we'll set up the development environment, which involves preparing the tools and resources needed to develop the project.

Setting up the development environment

With a theoretical understanding of TDD and its role in Angular application development, the next step is to configure our development environment so that we can apply TDD principles in Angular. To do this, we need to install the required tools to create an Angular project. By following the right steps, we can create a development environment that promotes effective and efficient development using TDD principles in Angular.

Setting up a development environment for Angular can be a bit more complex than setting up a general development environment. However, with the right guidance, it can be a straightforward process. Here, we will go over the steps you need to take to set up your Angular development environment. So, let's get started.

Installing Node.js on Windows or macOS

Follow these steps to install Node.js on Windows or macOS:

1. Go to the official Node.js website (`https://nodejs.org/en/`) and click on the **Download** button for the LTS version. This will download the latest version of Node.js for Windows or macOS:

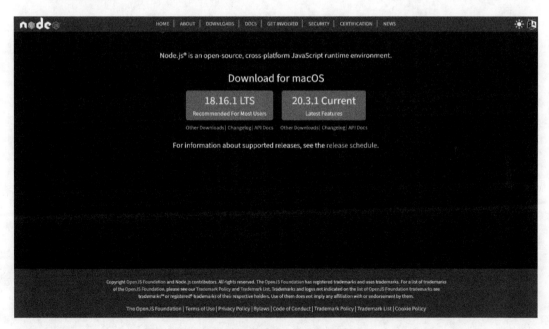

Figure 1.1 – Node.js home page

Once the installer has finished downloading, run it by double-clicking on the downloaded file. You should see the Node.js setup wizard:

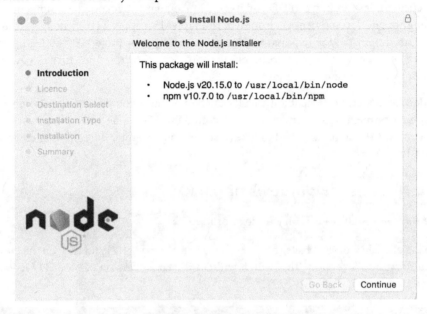

Figure 1.2 – Node.js installation – step 1

2. Read through the license agreement and click the **Agree** button if you agree to the terms:

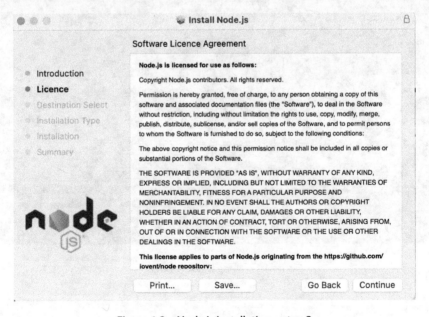

Figure 1.3 – Node.js installation – step 2

3. Next, choose the location where you want to install Node.js. The default location is usually fine, but you can choose a different location if you prefer:

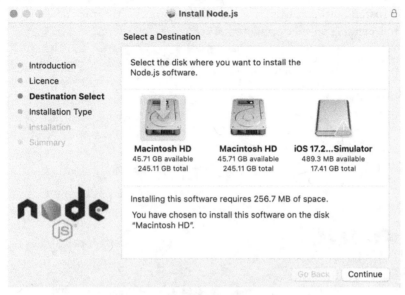

Figure 1.4 – Node.js installation – step 3

4. On the next screen, you'll be asked to choose which components to install. The default options are usually fine, but you can choose to add or remove components as needed:

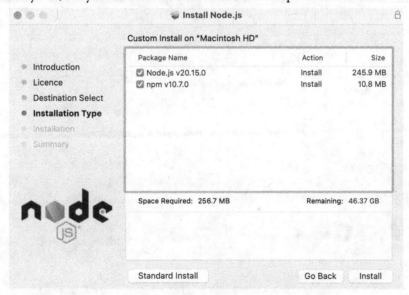

Figure 1.5 – Node.js installation – step 4

5. Choose the folder where you want to create the start menu shortcuts for Node.js. The default folder is usually fine, but you can choose a different folder if you prefer.

6. Click the **Install** button to begin the installation process. This may take a few minutes to complete, depending on your computer's speed:

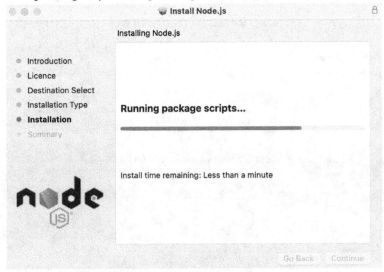

Figure 1.6 – Node.js installation – step 5

7. Once the installation is complete, you should see a message indicating that Node.js has been installed successfully. Click the **Close** button to close the installer:

Figure 1.7 – Node.js installation – step 6

8. To verify that Node.js has been installed correctly, open a command prompt and run the following:

    ```
    $ node -v
    ```

    This should display the version number of Node.js that you just installed:

Figure 1.8 – Checking the npm version

In the next section, we'll learn how to install Node.js on Linux.

Installing Node.js on Linux

Follow these steps to install Node.js on Linux:

1. Open your terminal and run the following command to update the package manager:

    ```
    $ sudo apt update
    ```

    The preceding command will give you the following output:

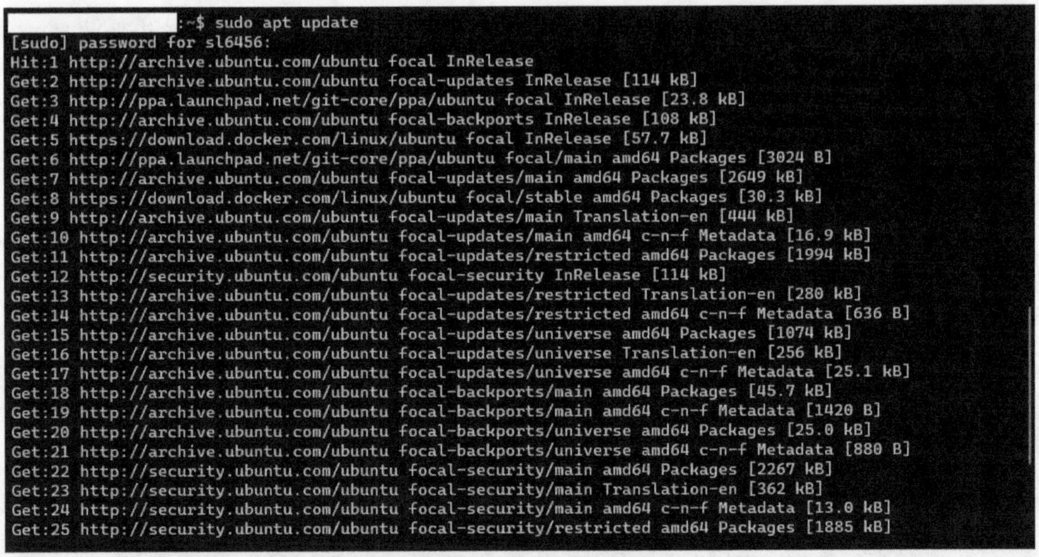

```
                   :~$ sudo apt update
[sudo] password for sl6456:
Hit:1 http://archive.ubuntu.com/ubuntu focal InRelease
Get:2 http://archive.ubuntu.com/ubuntu focal-updates InRelease [114 kB]
Get:3 http://ppa.launchpad.net/git-core/ppa/ubuntu focal InRelease [23.8 kB]
Get:4 http://archive.ubuntu.com/ubuntu focal-backports InRelease [108 kB]
Get:5 https://download.docker.com/linux/ubuntu focal InRelease [57.7 kB]
Get:6 http://ppa.launchpad.net/git-core/ppa/ubuntu focal/main amd64 Packages [3024 B]
Get:7 http://archive.ubuntu.com/ubuntu focal-updates/main amd64 Packages [2649 kB]
Get:8 https://download.docker.com/linux/ubuntu focal/stable amd64 Packages [30.3 kB]
Get:9 http://archive.ubuntu.com/ubuntu focal-updates/main Translation-en [444 kB]
Get:10 http://archive.ubuntu.com/ubuntu focal-updates/main amd64 c-n-f Metadata [16.9 kB]
Get:11 http://archive.ubuntu.com/ubuntu focal-updates/restricted amd64 Packages [1994 kB]
Get:12 http://security.ubuntu.com/ubuntu focal-security InRelease [114 kB]
Get:13 http://archive.ubuntu.com/ubuntu focal-updates/restricted Translation-en [280 kB]
Get:14 http://archive.ubuntu.com/ubuntu focal-updates/restricted amd64 c-n-f Metadata [636 B]
Get:15 http://archive.ubuntu.com/ubuntu focal-updates/universe amd64 Packages [1074 kB]
Get:16 http://archive.ubuntu.com/ubuntu focal-updates/universe Translation-en [256 kB]
Get:17 http://archive.ubuntu.com/ubuntu focal-updates/universe amd64 c-n-f Metadata [25.1 kB]
Get:18 http://archive.ubuntu.com/ubuntu focal-backports/main amd64 Packages [45.7 kB]
Get:19 http://archive.ubuntu.com/ubuntu focal-backports/main amd64 c-n-f Metadata [1420 B]
Get:20 http://archive.ubuntu.com/ubuntu focal-backports/universe amd64 Packages [25.0 kB]
Get:21 http://archive.ubuntu.com/ubuntu focal-backports/universe amd64 c-n-f Metadata [880 B]
Get:22 http://security.ubuntu.com/ubuntu focal-security/main amd64 Packages [2267 kB]
Get:23 http://security.ubuntu.com/ubuntu focal-security/main Translation-en [362 kB]
Get:24 http://security.ubuntu.com/ubuntu focal-security/main amd64 c-n-f Metadata [13.0 kB]
Get:25 http://security.ubuntu.com/ubuntu focal-security/restricted amd64 Packages [1885 kB]
```

Figure 1.9 – Updating Ubuntu packages

2. Run the following command to install Node.js:

```
$ sudo apt install nodejs
```

The preceding command will give you the following output:

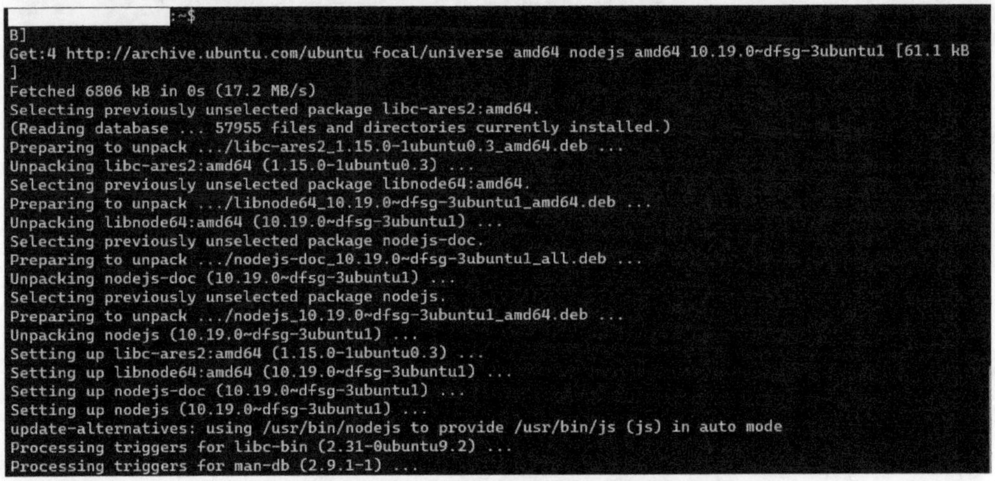

```
                   :~$
B]
Get:4 http://archive.ubuntu.com/ubuntu focal/universe amd64 nodejs amd64 10.19.0~dfsg-3ubuntu1 [61.1 kB
]
Fetched 6806 kB in 0s (17.2 MB/s)
Selecting previously unselected package libc-ares2:amd64.
(Reading database ... 57955 files and directories currently installed.)
Preparing to unpack .../libc-ares2_1.15.0-1ubuntu0.3_amd64.deb ...
Unpacking libc-ares2:amd64 (1.15.0-1ubuntu0.3) ...
Selecting previously unselected package libnode64:amd64.
Preparing to unpack .../libnode64_10.19.0~dfsg-3ubuntu1_amd64.deb ...
Unpacking libnode64:amd64 (10.19.0~dfsg-3ubuntu1) ...
Selecting previously unselected package nodejs-doc.
Preparing to unpack .../nodejs-doc_10.19.0~dfsg-3ubuntu1_all.deb ...
Unpacking nodejs-doc (10.19.0~dfsg-3ubuntu1) ...
Selecting previously unselected package nodejs.
Preparing to unpack .../nodejs_10.19.0~dfsg-3ubuntu1_amd64.deb ...
Unpacking nodejs (10.19.0~dfsg-3ubuntu1) ...
Setting up libc-ares2:amd64 (1.15.0-1ubuntu0.3) ...
Setting up libnode64:amd64 (10.19.0~dfsg-3ubuntu1) ...
Setting up nodejs-doc (10.19.0~dfsg-3ubuntu1) ...
Setting up nodejs (10.19.0~dfsg-3ubuntu1) ...
update-alternatives: using /usr/bin/nodejs to provide /usr/bin/js (js) in auto mode
Processing triggers for libc-bin (2.31-0ubuntu9.2) ...
Processing triggers for man-db (2.9.1-1) ...
```

Figure 1.10 – Node.js installation on Ubuntu

3. To verify that Node.js has been installed correctly, run the following command:

```
$ node -v
```

This should display the version number of Node.js that you just installed:

```
:~$ node -v
v18.16.1
```

Figure 1.11 – Node.js version after installing it on Ubuntu

4. npm is the package manager for Node.js. To install it, if you've not already installed it with Node.js, run the following command:

```
$ sudo apt install npm
```

Here's the output:

```
:~$ sudo apt install npm
[sudo] password for sl6456:
Reading package lists... Done
Building dependency tree
Reading state information... Done
The following additional packages will be installed:
  gyp libauthen-sasl-perl libdata-dump-perl libencode-locale-perl libfile-basedir-perl
  libfile-desktopentry-perl libfile-listing-perl libfile-mimeinfo-perl libfont-afm-perl
  libhtml-form-perl libhtml-format-perl libhtml-parser-perl libhtml-tagset-perl libhtml-tree-perl
  libhttp-cookies-perl libhttp-daemon-perl libhttp-date-perl libhttp-message-perl
  libhttp-negotiate-perl libio-html-perl libio-socket-ssl-perl libio-stringy-perl
  libipc-system-simple-perl libjs-inherits libjs-is-typedarray libjs-psl libjs-typedarray-to-buffer
  liblwp-mediatypes-perl liblwp-protocol-https-perl libmailtools-perl libnet-dbus-perl
  libnet-http-perl libnet-smtp-ssl-perl libnet-ssleay-perl libnode-dev libpython2-stdlib
  libpython2.7-minimal libpython2.7-stdlib libssl-dev libssl1.1 libtie-ixhash-perl libtimedate-perl
  libtry-tiny-perl liburi-perl libuv1-dev libwww-perl libwww-robotrules-perl libx11-protocol-perl
  libxml-parser-perl libxml-twig-perl libxml-xpathengine-perl node-abbrev node-ajv node-ansi
  node-ansi-align node-ansi-regex node-ansi-styles node-ansistyles node-aproba node-archy
  node-are-we-there-yet node-asap node-asn1 node-assert-plus node-asynckit node-aws-sign2 node-aws4
  node-balanced-match node-bcrypt-pbkdf node-bl node-bluebird node-boxen node-brace-expansion
  node-builtin-modules node-builtins node-cacache node-call-limit node-camelcase node-caseless
  node-chalk node-chownr node-ci-info node-cli-boxes node-cliui node-clone node-co node-color-convert
  node-color-name node-colors node-columnify node-combined-stream node-concat-map node-concat-stream

The following packages will be upgraded:
  libssl1.1
1 upgraded, 335 newly installed, 0 to remove and 241 not upgraded.
Need to get 13.6 MB of archives.
After this operation, 61.9 MB of additional disk space will be used.
Do you want to continue? [Y/n] |

  Setting up node-libnpx (10.2.1-2) ...
  Setting up npm (6.14.4+ds-1ubuntu2) ...
  Setting up libwww-perl (6.43-1) ...
  Setting up liblwp-protocol-https-perl (6.07-2ubuntu2) ...
  Setting up libxml-parser-perl (2.46-1) ...
  Setting up libxml-twig-perl (1:3.50-2) ...
  Setting up libnet-dbus-perl (1.2.0-1) ...
  Processing triggers for man-db (2.9.1-1) ...
  Processing triggers for mime-support (3.64ubuntu1) ...
  Processing triggers for libc-bin (2.31-0ubuntu9.2) ...
              :~$ |
```

Figure 1.12 – npm installation on Ubuntu

5.   To verify that npm has been installed correctly, run the following command:

```
$ npm -v
```

You should see the following output:

Figure 1.13 – npm version after Node.js installation

In the next section, we'll create a new Angular project.

Creating a new Angular project

With our development environment set up and ready, we can create our Angular project. This involves using the Angular CLI to generate the basic structure of our project, including files and folders. Once created, we can begin building our Angular application using the powerful tools and features provided by the framework.

After installing the Angular CLI, you can create a new Angular project by running the following command in your terminal:

```
$ ng new getting-started-angular-tdd --routing
```

This will create a new Angular project called getting-started-angular-tdd in the current directory.

After creating your Angular project, you can serve it by running the following command in your terminal:

```
$ cd getting-started-angular-tdd
$ ng serve --open
```

This will start a development server and open your application in your default browser. Here, you can make changes to your code and see the changes live in your browser.

Now, it's time to explore the different files involved in writing tests.

Exploring different files related to writing tests

Now that we've created the Angular project, we will explore the different files related to writing tests in Angular. I will provide insights into their role and best practices for working with them. Let's get started.

*.spec.ts files

`*.spec.ts` files contain the actual test cases that will be run against your code. These files are the backbone of testing in Angular as they define the individual test cases that will ensure your code works as expected. The name of the file should match the name of the file being tested, and it should be located in the same directory as the file being tested. The tests in these files are organized into test suites, which are defined using the `describe()` function. Each test case is defined using the `it()` function. For example, if you were testing a component named `MyComponent`, you would create a file named `my-component.spec.ts` and define the test cases for that component within that file.

The `describe()` function is used to group related test cases together, and it takes two parameters: a string that describes the test suite and a function that defines the test cases within that suite. The `it()` function is used to define individual test cases, and it takes two parameters: a string that describes the test case and a function that contains the code for the test case. Within the test case function, you can use the `expect()` function to define the expected behavior of your code. For example, you might use `expect(component.title).toEqual('My Title')` to test that the `title` property of a component has the expected value.

`*.spec.ts` files also typically import the component or service being tested, as well as any necessary dependencies. For example, if you were testing a component that used the `HttpClient` service, you would need to import both the component and `HttpClientTestingModule` from `@angular/common/http/testing`.

The karma.conf.js file

The `karma.conf.js` file is used to configure the Karma test runner, which is used to run your test cases. Karma is a popular test runner for Angular applications, and it provides a simple way to run your tests in a variety of browsers. The `karma.conf.js` file specifies the files that should be included in the test run, as well as the browsers that should be used to run the tests.

The `karma.conf.js` file typically exports a configuration object that specifies these settings. The configuration object contains several properties, such as frameworks, files, reporters, and browsers. These properties allow you to configure various aspects of the test run, such as which testing framework to use, which files to include in the test run, which reporters to use to display test results, and which browsers to use to run the tests.

For example, a typical `karma.conf.js` file might look like this:

```
module.exports = function(config) {
  config.set({
    frameworks: ['jasmine', '@angular/cli'],
    files: [
      { pattern: './src/test.ts', watched: false }
    ],
    reporters: ['progress', 'kjhtml'],
```

```
    browsers: ['Chrome']
  });
};
```

The given configuration outlines the utilization of the `Jasmine` and `@angular/cli` frameworks, the inclusion of the `./src/test.ts` file during the test run, the implementation of the `progress` and `kjhtml` reporters to showcase test results, and the execution of the tests in the Chrome browser.

The test.ts file

The `test.ts` file is the entry point for your test cases. It sets up the testing environment and loads all of the necessary files for the test run. This file typically imports the `zone.js` library, which is used to handle asynchronous operations in your tests. It also imports the Karma test runner and starts the test run.

The `test.ts` file is typically located in the `src` directory of your project, and it is usually generated automatically when you create a new Angular project. This file sets up the testing environment and loads the necessary files for the test run.

Here's an example of what a `test.ts` file might look like:

```
// This file is required by karma.conf.js and loads recursively all
the .spec and framework files
import 'zone.js/dist/zone-testing';
import { getTestBed } from '@angular/core/testing';
import { BrowserDynamicTestingModule, platformBrowserDynamicTesting }
from '@angular/platform-browser-dynamic/testing';

// First, initialize the Angular testing environment
getTestBed().initTestEnvironment(
  BrowserDynamicTestingModule,
  platformBrowserDynamicTesting()
);

// Then, load all the .spec files
const context = require.context('./', true, /\.spec\.ts$/);
context.keys().map(context);
```

This file initializes the Angular testing environment using the `getTestBed().initTestEnvironment()` function, which sets up `TestBed`. It also loads all of the `*.spec.ts` files using `require.context()`.

The tsconfig.spec.json file

The tsconfig.spec.json file is used to configure the TypeScript compiler for your test cases. It specifies the compiler options that should be used when compiling your test files. This file typically extends the main tsconfig.json file for your project but may include additional settings specific to testing.

Here's an example of what a tsconfig.spec.json file might look like:

```json
{
  "extends": "./tsconfig.json",
  "compilerOptions": {
    "outDir": "./out-tsc/spec",
    "module": "commonjs",
    "target": "es5",
    "baseUrl": "",
    "types": [
      "jasmine",
      "node"
    ]
  },
  "files": [
    "src/test.ts",
    "src/polyfills.ts"
  ],
  "include": [
    "**/*.spec.ts",
    "**/*.d.ts"
  ]
}
```

This file extends the main tsconfig.json file and specifies the compiler options that should be used for test files. It also includes the src/test.ts and src/polyfills.ts files in the test run.

The src/test.ts file

The src/test.ts file is used to configure the Angular testing environment. It sets up TestBed, which is used to create instances of your components and services for testing. It also imports any necessary testing utilities, such as TestBed and async.

Here's an example of what a `src/test.ts` file might look like:

```
import { TestBed } from '@angular/core/testing';
import {
BrowserDynamicTestingModule,  platformBrowserDynamicTestingModule }
from '@angular/platform-browser-dynamic/testing';

TestBed.initTestEnvironment(
  BrowserDynamicTestingModule,
  platformBrowserDynamicTestingModule()
);
```

This file sets up `TestBed` using the `TestBed.initTestEnvironment()` function. It specifies the testing module and platform module to use, which are the `BrowserDynamicTestingModule` and `platformBrowserDynamicTestingModule`, respectively.

Summary

This chapter covered the basics of TDD and its role in Angular. We explained the benefits of TDD and how it can help developers write high-quality code. Then, we discussed how to set up the development environment for Angular and created a new Angular project using the Angular programming interface. We also explored the various files related to writing tests in Angular, including `*.spec.ts`, `karma.conf.js`, `tsconfig.spec.json`, and `src/test.ts`. We provided detailed explanations of each file and their role in testing Angular applications. By understanding these files and their purpose, developers can write and run tests for their Angular applications more effectively and ensure that their code performs as expected.

In the next chapter, we'll learn about the process of writing and executing unit tests using Jasmine while covering topics such as test suites, test specifications, and matchers.

2

Using Jasmine and Karma to Test Angular Applications

Jasmine and Karma are two powerful tools that developers can use to test their Angular applications. Testing is an essential part of the development process as it helps ensure that the application works as expected and avoids any potential bugs or issues.

Jasmine is a **behavior-driven development (BDD)** framework for testing JavaScript code. It provides a simple and readable syntax for writing tests, making it easier to understand and maintain the code. With Jasmine, developers can define test suites and test cases, and then use various matchers to check the expected behavior of their code.

Karma, on the other hand, is a test runner that allows developers to execute their tests in multiple browsers and environments. It provides a seamless integration with Jasmine, allowing developers to easily run their Jasmine tests in different browsers and get real-time feedback on the test results. Karma also offers additional features, such as code coverage reporting and continuous integration support.

Using Jasmine and Karma together can greatly enhance the testing process for Angular applications. Developers can write comprehensive test suites using Jasmine's expressive syntax, and then use Karma to run these tests in various browsers, ensuring compatibility across different environments. This helps catch any potential issues or bugs early on and promotes a more robust and reliable application.

In this chapter, we will look into the basics of using Jasmine and Karma for testing Angular applications. We will learn how to set up the testing environment, write unit tests with Jasmine, and configure Karma to run the tests in different browsers.

The following topics will be covered in this chapter:

- Mastering Jasmine's unit testing techniques
- Writing your first unit tests in Angular related to **test-driven development (TDD)**
- Utilizing code coverage and test result analysis with Karma

Technical requirements

To follow along with the examples and exercises in this chapter, you will need to have a basic understanding of Angular and TypeScript, as well as the following technical requirements:

- Node.js LTS and npm LTS installed on your computer
- Angular 17 or later CLI installed globally
- A code editor, such as Visual Studio Code, installed on your computer

The code files of this chapter can be found at `https://github.com/PacktPublishing/ Mastering-Angular-Test-Driven-Development/tree/main/Chapter%202`.

Mastering Jasmine's unit testing techniques

In this section, we'll explore the Jasmine framework by writing descriptive test suites, utilizing matchers, using spies for function testing, and testing asynchronous code. By leveraging these techniques, you can ensure the quality and reliability of your code base.

What is Jasmine?

Jasmine is a widely used testing framework for JavaScript that's commonly employed for writing tests for web applications and Node.js projects. With its clean and expressive syntax, Jasmine allows developers to create easy-to-understand and maintainable test suites and cases. It offers built-in functionalities for assertions, test spies, and asynchronous testing. Jasmine seamlessly integrates with numerous libraries and frameworks, including AngularJS and React, enabling developers to compose comprehensive and dependable tests for JavaScript applications. Its popularity stems from its simplicity, flexibility, and ability to facilitate the creation of robust and reliable tests for JavaScript code.

With Jasmine, developers can structure their tests using a BDD style, making it easy to write tests that are both descriptive and readable. It provides a set of built-in functions for assertions, which allow developers to verify the expected behavior of their code. These assertions cover a wide range of scenarios and make it simple to write tests that validate the correctness of the code being tested.

Jasmine also includes features such as test spies, which enable developers to track function calls and arguments, as well as mock and stub function behavior. This helps in testing code that interacts with other components or external dependencies.

Furthermore, Jasmine supports asynchronous testing, making it easy to write tests for code that involves asynchronous operations such as AJAX requests or timers. It provides mechanisms to handle asynchronous tasks and ensure that tests wait for completion before making assertions.

Jasmine is highly extensible and can be used in conjunction with various libraries and frameworks, such as AngularJS, Angular, and React. It integrates seamlessly with these ecosystems, allowing developers to write comprehensive and reliable tests for their JavaScript applications.

Overall, Jasmine's simplicity, flexibility, and comprehensive feature set have contributed to its popularity as a testing framework for JavaScript. It empowers developers to write robust and reliable tests, ultimately leading to higher-quality code and more confident deployments.

Writing descriptive test suites

One of the fundamental principles of effective unit testing is writing descriptive test suites. By logically organizing your tests and using descriptive names, you make it easier for yourself and other developers to understand the purpose and behavior of each test. In this section, we will explore strategies for creating meaningful test suite names and describing the expected behavior in clear and concise language. Additionally, we will discuss how descriptive test suites can serve as documentation for future reference.

A descriptive test suite is a collection of related test cases that focuses on a specific functionality or component of your code. It serves as a documentation tool and helps developers understand the purpose and behavior of each test. Descriptive test suites are essential for maintaining code quality, facilitating collaboration among team members, and ensuring that tests remain relevant and up-to-date over time. By investing time in creating descriptive test suites, you can improve the maintainability and readability of your test code.

Let's consider a simple scenario where we have a JavaScript function called `calculateTotal` that calculates the total price of items in a shopping cart. We want to write a test to ensure that the function returns the correct total when given a set of items with their respective prices.

Choosing meaningful names

The first step in creating descriptive test suites is choosing meaningful names for your test suites and test cases. Use clear and concise language to describe the functionality or behavior being tested. Avoid ambiguous or generic names that don't provide enough context. For example, instead of naming a test suite "Test Suite 1," consider naming it "User Authentication Tests" to convey the purpose of the tests. Meaningful names make it easier for developers to locate specific tests and understand their purpose, even when revisiting the code base after a long time.

Structuring test suites

Organizing your test suites in a logical and hierarchical structure is crucial for creating descriptive test suites in Jasmine. A well-structured test suite mirrors the structure of your code base, making it easier to locate and understand specific tests. Group related tests together to improve readability and maintainability. For example, if you are testing a user authentication module, create a test suite specifically for login functionality and another for registration. This separation helps you isolate and focus on specific features, making it easier to identify and resolve issues. Additionally, consider using nested describe blocks to further organize your tests hierarchically. Here's an example:

```
describe("User Authentication", () => {
  describe("Login", () => {
    // Login-related test cases
  });

  describe("Registration", () => {
    // Registration-related test cases
  });
});
```

Writing clear and concise test descriptions

Within each test case, write clear and concise descriptions that accurately describe the expected behavior. Use language that is easily understandable and avoids technical jargon whenever possible. A well-written test description should provide enough information for you and others to understand the purpose of the test without needing to dive into the implementation details. Consider using the "should" format to describe the expected behavior – for example, "should correctly calculate the total for a cart with multiple items." By using descriptive language, future developers can quickly grasp the intent of the test and identify any deviations from the expected behavior.

In addition to the test description, it is also helpful to include comments within the test code to provide further clarification or context where needed. These comments can explain the reasoning behind certain assertions or provide additional information about the test scenario. However, it is important to strike a balance and avoid excessive commenting that may clutter the test code.

Maintaining and updating descriptive test suites

Descriptive test suites are not a one-time effort but require ongoing maintenance and updates as the code base evolves. It is essential to review and update test suites regularly to ensure they remain relevant and accurate. When making changes to the code, developers should also update the corresponding tests to reflect the updated behavior. Additionally, if a test case becomes obsolete or redundant, it should be removed or refactored.

When updating test suites, it is crucial to keep their descriptive nature intact. If a test case needs significant changes, it may be beneficial to create a new test case with an appropriate description instead of modifying the existing one. This helps maintain the clarity and transparency of the test suite.

Let's consider a simple scenario where we have a JavaScript function called `calculateTotal` that calculates the total price of items in a shopping cart. We want to write a test to ensure that the function returns the correct total when given a set of items with their respective prices:

```javascript
// Function under test
function calculateTotal(items) {
  let total = 0;
  items.forEach(item => {
    total += item.price;
  });
  return total;
}


// Test suite
describe("calculateTotal function", () => {
  // Test case 1: Calculate total for an empty cart
  it("should return 0 for an empty cart", () => {
    const cart = [];
    const result = calculateTotal(cart);
    expect(result).toBe(0);
  });

  // Test case 2: Calculate total for a cart with multiple items
  it("should correctly calculate the total for a cart with multiple
items", () => {
    const cart = [
      { name: "Item 1", price: 10 },
      { name: "Item 2", price: 15 },
      { name: "Item 3", price: 20 }
    ];
    const result = calculateTotal(cart);
    expect(result).toBe(45);
  });
});
```

In the preceding example, we have created a test suite for the `calculateTotal` function. Within the test suite, we have two test cases, and the descriptions of the test cases clearly state what behavior is being tested:

- The first test case, *"should return 0 for an empty cart,"* verifies that the function correctly handles an empty shopping cart and returns a total of 0

- The second test case, *"should correctly calculate the total for a cart with multiple items,"* tests the function with a cart containing multiple items and checks if the calculated total is as expected

By providing descriptive test case descriptions, other developers can easily understand the intent and behavior of each test. These descriptions act as documentation, making it easier to maintain and update the tests as the code base evolves.

In the next section, we'll look at how to write our first unit tests using TDD principles.

Writing your first unit tests in an Angular project

Unit testing is a critical aspect of Angular development that ensures code quality, reliability, and maintainability. TDD is a software development approach that emphasizes writing tests before implementing the actual code. In this section, you'll learn how to write your first unit tests in an Angular project while following the principles of TDD. By leveraging the Jasmine testing framework and Angular's testing utilities, developers can create effective and robust unit tests that verify the correctness of their code.

We'll be using the project we created in *Chapter 1* to practice. Follow these steps to write your first unit test:

1. Create a new component called `CalculatorComponent` by running the following command:

```
$ ng g m calculator --route calculator --module=app.module
```

After creating the component with the preceding command line, a `calculator.component.spec.ts` file will be created in the `src/ app/calculator` folder. When you open the file, you'll see the following code by default:

```
import { ComponentFixture, TestBed } from '@angular/core/
testing';

import { CalculatorComponent } from './calculator.component';

describe('CalculatorComponent', () => {

let calculator: CalculatorComponent;

let fixture: ComponentFixture<CalculatorComponent>;
```

```
beforeEach(async () => {

await TestBed.configureTestingModule({

declarations: [ CalculatorComponent ]

})

.compileComponents();

fixture = TestBed.createComponent(CalculatorComponent);

calculator = fixture.componentInstance;
fixture.detectChanges();

});

it('should create', () => {

expect(calculator).toBeTruthy();

});

});
```

In the preceding generated code, we have a test suite where we have used the `describe` function, providing a descriptive name for the component under test. Within the test suite, we have a `beforeEach` block to set up the test environment. The `TestBed.configureTestingModule` method is used to configure the test module and provide the necessary dependencies. The `calculator` variable is then assigned to an instance of `CalculatorComponent` using the `TestBed.inject` method.

Our `CalculatorComponent` component will enable us to perform basic arithmetic operations. To write a unit test using TDD, we'll start by creating a test case that verifies the component's expected behavior.

2. Now, we'll write the actual test case using the `it` function. In this case, we'll test the `add` method of `CalculatorComponent` by passing it two numbers and expecting the result to be 5. The `expect` function is used to define the expected behavior and check the actual result. The following code must be added to the test suite – that is, inside the `describe` function:

```
it('should add two numbers correctly', () => {
    const result = calculator.add(2, 3);
    expect(result).toBe(5);
```

```
    });
  });
```

3. You will get an error in your code editor telling you that the `add` function doesn't exist:

```
it('should add two numbers correctly', () ⇒ {
    const result = calculator.add(2, 3);
    expect(result).toBe(5);
});
```

Figure 2.1 – Code error

This is normal as it hasn't been created yet.

Upon returning to our Karma server, we'll see that our test case isn't displayed in `CalculatorComponent` and that in the terminal, we have an error related to the non-existence of the function and a message indicating that no test has succeeded.

Don't panic – it's the red of TDD! Well done!

4. Next, we will implement the `add` function in `calculator.component.ts`. Having defined our first test case, we can proceed to implement `calculator.component.ts` for the test to pass. Following the TDD approach, write the minimum amount of code necessary to pass the tests:

```
add(a: number, b: number): number {
    return a + b;
}
```

You'll see the following result on your Karma server:

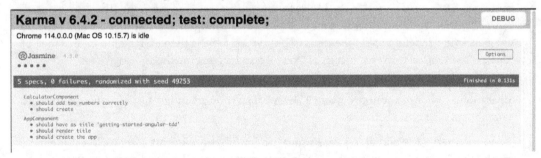

Figure 2.2 – The test succeeded

In your terminal, you will receive the following message:

```
✓ Browser application bundle generation complete.
Chrome 114.0.0.0 (Mac OS 10.15.7): Executed 5 of 5 SUCCESS (0.131 secs / 0.119 secs)
TOTAL: 5 SUCCESS
```

Figure 2.3 – The test was executed successfully

> With that, we're at the green stage of TDD and writing the minimum amount of code needed for our test to pass. Well done!

Once the tests have passed, you can refactor the code to improve its design, readability, and maintainability. Refactoring is an essential step in the TDD process as it helps eliminate duplication and improves code structure and overall quality. It is essential to ensure that tests continue to run after refactoring. Regularly reviewing and updating tests as the code base evolves will help maintain the integrity and reliability of unit tests.

In our example, we don't need to refactor the test. Don't worry – we'll have the opportunity to do so in *Chapter 3*.

In the next section, we'll look at how to use code coverage and test result analysis with Karma.

Utilizing code coverage and test result analysis with Karma

Code coverage and test result analysis are essential aspects of the software development process. By measuring code coverage, developers can assess the effectiveness of their unit tests and identify areas that require additional testing. Karma, a popular testing framework in the JavaScript ecosystem, provides built-in support for code coverage and test result analysis. In this section, we'll learn how to leverage Karma to measure code coverage, generate detailed reports, and analyze test results. By utilizing these features, developers can ensure comprehensive testing and improve the overall quality and reliability of their code.

Before going any further, it's worth noting that all the configurations we'll be looking at are already present in our Angular project. When we create our project, Angular takes care of all the configuration.

Here are the different steps Angular performs for us:

- **Step 1 – setting up Karma with code coverage**:

  To utilize code coverage with Karma, start by installing the necessary dependencies:

  ```
  $ npm install --save-dev karma-coverage
  ```

  Next, configure Karma to generate code coverage reports. Update your Karma configuration file (`karma.conf.js`) with the following changes:

  ```
  module.exports = function(config) {
    config.set({
      // ...
      reporters: ['progress', 'coverage'],
      coverageReporter: {
        dir: 'coverage/',
        reporters: [
          { type: 'html', subdir: 'report-html' },
  ```

```
           { type: 'lcov', subdir: 'report-lcov' }
        ]
      },
      // ...
    });
  };
```

This configuration specifies the reporters to be used (`progress` for test progress and `coverage` for code coverage). The `coverageReporter` section defines the output directory and the types of reports to generate (HTML and LCOV).

- **Step 2 – running tests and generating code coverage reports**:

After configuring Karma for code coverage, run your tests as usual. Karma will now generate code coverage reports alongside the test results. These coverage reports provide insights into which parts of your code base are covered by tests and which areas require additional testing.

Once the tests have finished running, navigate to the `coverage` directory to view the generated reports. Open the HTML report (`coverage/report-html/index.html`) in a web browser to visualize the code coverage details. The report highlights covered lines, uncovered lines, and overall coverage percentages. Additionally, the LCOV report (`coverage/report-lcov/lcov-report/index.html`) provides a more detailed breakdown of code coverage.

- **Step 3 – analyzing the test results**:

Karma also offers features to analyze the test results, including test reporting and integration with popular **continuous integration** (**CI**) tools. By leveraging these capabilities, developers can gain insights into test failures, identify patterns, and track the overall health of their test suite.

Karma provides various reporters that offer different levels of detail in the test output. For example, `mocha-reporter` displays detailed information about test failures, including stack traces and error messages, whereas `junit-reporter` generates JUnit-style XML reports that can be consumed by CI tools for further analysis.

To integrate Karma with CI tools, configure the respective plugin or reporter in your Karma configuration file. For example, to generate JUnit reports for Jenkins, add the `karma-junit-reporter` plugin and configure it accordingly.

- **Step 4 – utilizing thresholds and quality gates**:

Karma allows developers to define thresholds and quality gates for code coverage and test results. By setting these thresholds, developers can establish the minimum requirements for code coverage and test success rates. This ensures that the code base maintains a certain level of quality and reduces the risk of shipping untested or poorly covered code.

To set thresholds for code coverage, update your Karma configuration file as follows:

```
module.exports = function(config) {
  config.set({
    // ...
    coverageReporter: {
      // ...
      check: {
        global: {
          statements: 80,
          branches: 80,
          functions: 80,
          lines: 80
        }
      }
    },
    // ...
  });
};
```

In this example, the thresholds have been set to 80% for statements, branches, functions, and lines. If any of these thresholds are not met, Karma will report a failed test result.

Code coverage visualization

In our project, we started by writing tests on `CalculatorComponent`. Now, we can see the code coverage using Karma. Let's run the following command in our project's terminal:

```
$ ng test –code-coverage
```

After executing the preceding command, we'll observe the following three things:

- In the terminal, we'll have the following, if all goes well:

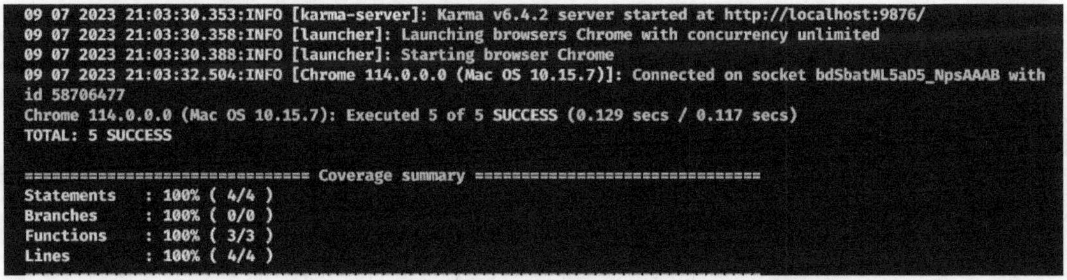

```
09 07 2023 21:03:30.353:INFO [karma-server]: Karma v6.4.2 server started at http://localhost:9876/
09 07 2023 21:03:30.358:INFO [launcher]: Launching browsers Chrome with concurrency unlimited
09 07 2023 21:03:30.388:INFO [launcher]: Starting browser Chrome
09 07 2023 21:03:32.504:INFO [Chrome 114.0.0.0 (Mac OS 10.15.7)]: Connected on socket bdSbatML5aD5_NpsAAAB with
id 58706477
Chrome 114.0.0.0 (Mac OS 10.15.7): Executed 5 of 5 SUCCESS (0.129 secs / 0.117 secs)
TOTAL: 5 SUCCESS

=============================== Coverage summary ===============================
Statements   : 100% ( 4/4 )
Branches     : 100% ( 0/0 )
Functions    : 100% ( 3/3 )
Lines        : 100% ( 4/4 )
```

Figure 2.4 – Test coverage in the terminal

- Karma launches our browser, showing us the various tests that were carried out:

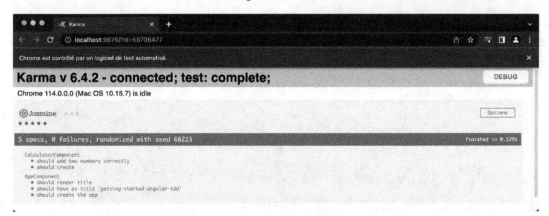

Figure 2.5 – A successful test

- A `coverage` folder is created in our project's arborescence:

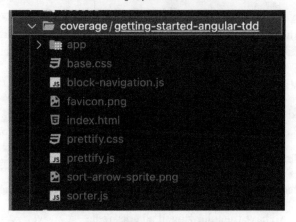

Figure 2.6 – The test coverage folder

Inside is an `index.html` file. When we launch the file in the browser, we'll see a table summarizing all the files tested, and in each file, we're told how much of a given piece of code has been tested.

The following screenshots show test coverage:

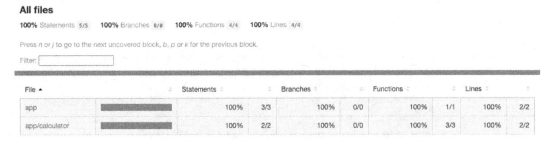

Figure 2.7 – Test coverage visualization on the web – part 1

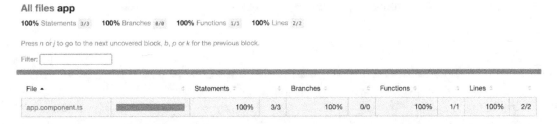

Figure 2.8 – Test coverage visualization on the web – part 2

All files / app **app.component.ts**

100% Statements 3/3   **100%** Branches 0/0   **100%** Functions 1/1   **100%** Lines 2/2

Press *n* or *j* to go to the next uncovered block, *b*, *p* or *k* for the previous block.

```
1       import { Component } from '@angular/core';
2
3       @Component({
4         selector: 'app-root',
5         templateUrl: './app.component.html',
6         styleUrls: ['./app.component.scss']
7       })
8   1x  export class AppComponent {
9   3x    title = 'getting-started-angular-tdd';
10      }
11
```

Figure 2.9 – Test coverage visualization on the web – part 3

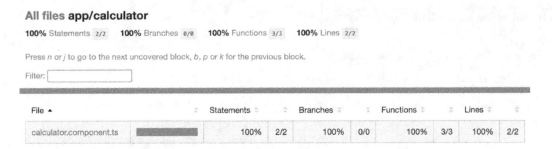

Figure 2.10 – Test coverage visualization on the web – part 4

All files / app/calculator calculator.component.ts

100% Statements 2/2   **100%** Branches 0/0   **100%** Functions 3/3   **100%** Lines 2/2

Press *n* or *j* to go to the next uncovered block, *b*, *p* or *k* for the previous block.

```typescript
1     import { Component, OnInit } from '@angular/core';
2
3     @Component({
4       selector: 'app-calculator',
5       templateUrl: './calculator.component.html',
6       styleUrls: ['./calculator.component.scss'],
7     })
8  1x  export class CalculatorComponent implements OnInit {
9       constructor() {}
10
11      ngOnInit(): void {}
12
13      add(a: number, b: number): number {
14 1x     return a + b;
15      }
16    }
17
```

Figure 2.11 – Test coverage visualization on the web – part 5

By utilizing code coverage and test result analysis with Karma, developers can enhance their testing practices and ensure comprehensive code coverage. Karma's built-in support for code coverage enables developers to measure the effectiveness of their tests and identify areas that require additional attention. Additionally, Karma's test reporting and integration capabilities allow for deeper analysis of test results, enabling developers to track the health of their test suite and identify patterns in test failures. By setting thresholds and quality gates, developers can establish minimum requirements for code coverage and test success rates, ensuring a higher level of code quality and reliability.

Summary

This chapter covered how to set up the testing environment, write unit tests with Jasmine, and configure Karma to run the tests in different browsers. Jasmine and Karma are powerful tools for testing Angular applications. Jasmine is a BDD framework that provides an intuitive syntax for writing test cases. Karma, on the other hand, is a test runner that allows you to execute tests in various environments and provides features such as code coverage and test result analysis.

To use Jasmine and Karma to test Angular applications, you need to set up the testing environment by installing the necessary dependencies and configuring Karma. Jasmine provides a rich set of matchers and assertions to validate the behavior of Angular components, services, and directives. You can create test suites and test cases to cover different scenarios and expectations.

Karma allows you to run tests in real browsers or headless environments, making it easy to simulate user interactions and test the application's behavior across different platforms. It also provides support for code coverage, generating reports that help identify areas of the code base that require additional testing.

By using Jasmine and Karma together, you can write complete unit tests for your Angular applications and practice the principles of TDD.

In the next chapter, we'll learn how to write effective unit tests for Angular components, services, and directives.

Part 2: Writing Effective Unit Tests

In this part, you'll write unit tests for components, services, and directives, using pipes, forms, and reactive programming while respecting TDD principles.

This part has the following chapters:

- *Chapter 3, Writing Effective Unit Tests for Angular Components, Services, and Directives*
- *Chapter 4, Mocking and Stubbing Dependencies in Angular Tests*
- *Chapter 5, Testing Angular Pipes, Forms, and Reactive Programming*

3

Writing Effective Unit Tests for Angular Components, Services, and Directives

In this chapter, we're going to dive into the art of writing effective unit tests for Angular components, services, and directives. We'll dive into writing unit tests for Angular components, continuing what we started in the previous chapter. Components are the building blocks of Angular applications, and it's crucial to test them thoroughly. In this chapter, we'll learn how to set up test environments, create component instances, and test component properties, methods, and event handling. We'll also explore techniques for testing component models, including DOM manipulation and event simulation.

We'll also be focusing on unit testing Angular services. Services play an essential role in Angular applications, providing reusable logic and data manipulation. We'll learn how to create service instances, simulate test service methods, and carry out data manipulation.

Finally, we'll look at unit testing Angular directives. Directives are powerful tools for manipulating the DOM and improving the behavior of our applications. We'll learn how to set up test environments for directives, create directive instances, and test their behavior and interaction with the DOM.

Throughout this chapter, I'll provide practical examples and real-world scenarios to illustrate effective unit-testing concepts and techniques. We'll also discuss common unit-testing pitfalls and challenges and provide strategies for overcoming them.

In summary, here are the main topics that will be covered:

- Advanced techniques for Angular unit testing: lifecycle hooks and dependencies

- Advanced techniques for Angular unit testing: Angular services

- Using rigorous directive testing to ensure proper rendering and functionality

By the end of this chapter, you'll have a solid understanding of how to write effective unit tests for Angular components, services, and directives. You'll be equipped with the knowledge and tools you need to ensure the quality and reliability of your Angular code. So, let's dive in and master the art of writing effective unit tests for Angular applications.

Technical requirements

To follow along with the examples and exercises in this chapter, you will need to have a basic understanding of Angular and TypeScript, as well as the following:

- Node.js and npm installed on your computer
- Angular CLI installed globally
- A code editor, such as Visual Studio Code, installed on your computer

The code files of this chapter can be found at `https://github.com/PacktPublishing/Mastering-Angular-Test-Driven-Development/tree/main/Chapter%203`.

Advanced techniques for Angular unit testing – lifecycle hooks

In this section, we will understand how to leverage lifecycle hooks and manage dependencies in our unit tests for Angular components. You'll be empowered with the knowledge and skills you need to write robust and efficient unit tests, ensuring the quality and stability of your Angular applications. Let's dive in and explore advanced unit-testing techniques for Angular components.

Discovering lifecycle hooks

Angular provides several lifecycle hooks that allow us to perform actions at specific stages of a component's lifecycle. Testing these hooks ensures that our components behave as expected. But before we look deeper into the subject of the testing lifecycle, let's take a look at some of Angular's lifecycle methods:

- `ngOnInit()`: The `ngOnInit()` hook is called after the component has been initialized. In our `Calculator` component, we can use this hook to set the initial values and perform any necessary setup. To test `ngOnInit()`, we can verify whether the initial values are correctly set and whether any necessary setup is performed.
- `ngOnChanges()`: The `ngOnChanges()` hook is called whenever there are changes to the component's input properties. In our `Calculator` component, we can use this hook to update the component state based on the changes. To test `ngOnChanges()`, we can simulate changes to the input properties and verify whether the component state is updated accordingly.

- ngOnDestroy(): The ngOnDestroy() hook is called just before the component is destroyed. In our Calculator component, we can use this hook to clean up any resources or subscriptions. To test ngOnDestroy(), we can simulate the component destruction and verify whether the necessary cleanup actions are performed.

In the next section, we will learn how to test the dependencies present in an Angular component.

Practical application

We will continue our project from the previous chapter, namely the Calculator application. We'll start testing the expected behavior when we initialize our calculator. When our calculator is launched, the result displayed should be 0, since no operation has been performed. To do this, we'll test the ngOnInit() method of the Angular lifecycle, which allows us to initialize our component.

In calculator.component.spec.ts, we will add the following unit test:

```
it('should initialize result to 0', () => {
    calculator.ngOnInit();
    expect(calculator.result).toEqual(0);
  });
```

After writing this code snippet, you'll notice that there are errors in the code because result is not an attribute of the Calculator class:

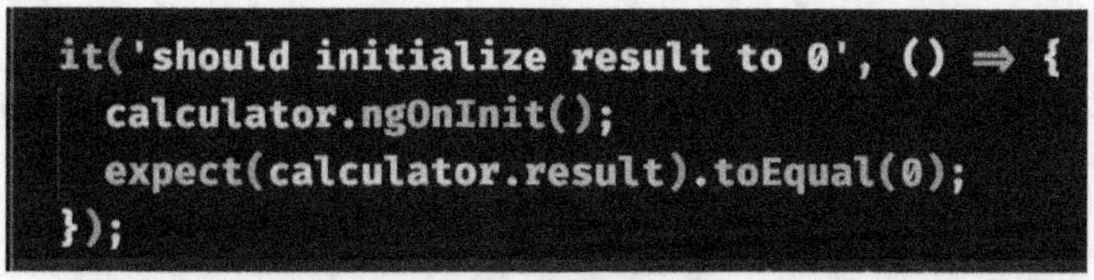

```
it('should initialize result to 0', () => {
  calculator.ngOnInit();
  expect(calculator.result).toEqual(0);
});
```

Figure 3.1 – ngOnInit method test case with error

Don't forget the principles of TDD. It's normal for the test to fail in the first instance and then for you to write the minimum amount of code necessary for the test to succeed. In the meantime, we need to rectify the problem with our result class attribute by declaring it in our component.

Our component should now look like this:

```typescript
import { Component, OnInit } from '@angular/core';

@Component({
  selector: 'app-calculator',
  templateUrl: './calculator.component.html',
  styleUrls: ['./calculator.component.scss'],
})
export class CalculatorComponent implements OnInit {
  result!: number;

  constructor() {}

  ngOnInit(): void {}

  add(a: number, b: number): number {
    return a + b;
  }
}
```

Figure 3.2 – Adding the add method in calculator.component.ts

Now that the declaration issue has been resolved, let's focus on our red test. We have an error telling us that the result should be initialized to 0:

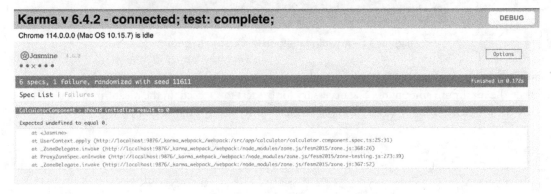

Figure 3.3 – calculator.component.ts test failed

To fix this, we just need to initialize the `result` value to 0 in our `ngOnInit()` method:

```
ngOnInit(): void {
   this.result = 0;
}
```

As a result, our test turns green. Well done!

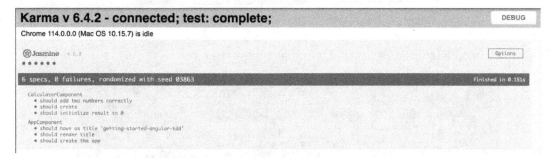

Karma v 6.4.2 - connected; test: complete;   DEBUG

Chrome 114.0.0.0 (Mac OS 10.15.7) is idle

Jasmine 4.6.2   Options

● ● ● ● ● ●

6 specs, 0 failures, randomized with seed 03863   finished in 0.151s

CalculatorComponent
   ● should add two numbers correctly
   ● should create
   ● should initialize result to 0

AppComponent
   ● should have as title 'getting-started-angular-tdd'
   ● should render title
   ● should create the app

Figure 3.4 – calculator.component.ts test succeeded

We've now tested the `ngOnDestroy()` method of our component's lifecycle, ensuring that our component is initialized with the expected values, while respecting the principles of TDD.

This is the same philosophy to be adopted for other lifecycle methods. To be able to test our `ngOnDestroy()` method, we add a small layer to our business logic.

Let's assume the following scenario.

We have a service that handles the various arithmetic operations of our calculator and returns the result to us. This service is injected into our `CalculatorComponent` to call the various methods that would return the result following the arithmetic operation.

As we are used to by now, we will first initialize our `CalculatorService` in the test file of our `CalculatorComponent`:

```
Complexity is 5 Everything is cool!
describe('CalculatorComponent', () => {
   let calculator: CalculatorComponent;
   let fixture: ComponentFixture<CalculatorComponent>;
   let calculatorService: CalculatorService;
```

Figure 3.5 – Declaration of an instance of CalculatorService

Then, since it's a service, we need to declare it in the `providers` array:

```
beforeEach(async () ⇒ {
  await TestBed.configureTestingModule({
    declarations: [CalculatorComponent],
    providers: [CalculatorService]
  }).compileComponents();
```

Figure 3.6 – Adding CalculatorService in the providers array

Finally, we'll inject it so that it can be used in our descriptive test suites:

```
beforeEach(async () ⇒ {
  await TestBed.configureTestingModule({
    declarations: [CalculatorComponent],
    providers: [CalculatorService]
  }).compileComponents();

  fixture = TestBed.createComponent(CalculatorComponent);
  calculator = fixture.componentInstance;
  calculatorService = TestBed.inject(CalculatorService);
  fixture.detectChanges();
});
```

Figure 3.7 – Injection of CalculatorService in our test context

We can now create our service so that it is recognized by our test file. In our project's `src` folder, create a `core` folder containing a `services` folder. Basically, you'll have `src/core/services`.

Open the `services` folder in the terminal and run the following command:

```
ng g s calculator
```

Once the service has been created, import it into the test file and the errors will disappear from your code editor:

```
import { CalculatorService } from 'src/core/services/calculator.
service';
```

According to our scenario, the service is now in charge of performing arithmetic operations. So, we'll move the logic of our `add()` method from the component to the service.

Here's how it would look on the test side:

```
it('should add two numbers correctly', () => {
  spyOn(calculatorService, 'add').and.callThrough();
  calculator.add(2, 3);
  expect(calculatorService.add).toHaveBeenCalledWith(2, 3);
  expect(calculator.result).toBe(5);
});
```

Figure 3.8 – Add method test case

We've refactored our `should add two numbers correctly` test suite. Previously, the calculation was performed directly in the component. Now, it's transferred to a service that takes care of it. After calculation, the service returns the result to the component.

The service's `add()` method must perform the sum and return the result.

The component's `add()` method calls on the service's `add()` method to retrieve the result.

Now we just need to declare the `add()` method in our service, so that our editor code no longer contains errors:

```
import { Injectable } from '@angular/core';

@Injectable({
  providedIn: 'root'
})
export class CalculatorService {

  constructor() { }

  add(a: number, b: number) {}
}
```

Figure 3.9 – Adding an add method declaration in calculator.service.ts

Our test now compiles and displays an error:

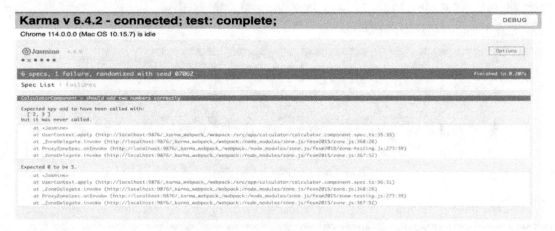

Figure 3.10 – add method test case failed

We can now use the minimum amount of code necessary to turn our service green. To do this, our service's `add()` method must return a number (we'll limit ourselves to an integer). It then takes two arguments, namely the numbers a and b:

```
add(a: number, b: number): number {}
```

Finally, it must return the sum of a and b:

```
add(a: number, b: number): number {
  return a + b;
}
```

Now that the work is done on the service side, we need to update our component code. First, we'll inject the service into our component's constructor:

```
constructor(private calculatorService: CalculatorService) {}
```

Then, our component's `add()` method becomes this:

```
add(a: number, b: number): void {
  this.result = this.calculatorService.add(a, b);
}
```

If you notice, we've gone from a function that returned a `number` value to a `void`. The `result` value is returned directly by our service. If all goes well, the `ng test` command returns the following on the screen:

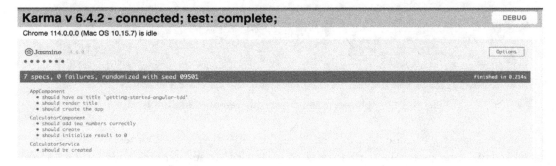

Figure 3.11 – calculator.component.ts test cases succeeded

We've just carried out a small dependency test using our service. In the next section, we'll go a little further in relation to the tests related to our service.

Advanced techniques for Angular unit testing – Angular services

In this section, we will explore advanced techniques for unit testing Angular services. We will dive into the various aspects of testing services, including testing methods, HTTP requests, observables, and error handling. By mastering these techniques, you will be able to write comprehensive and robust unit tests for your Angular services, ensuring that they perform as expected and handle various scenarios gracefully.

Testing service methods

Services typically contain methods that perform specific operations or logic. These methods can be tested individually to ensure they produce the expected results. By mocking any dependencies and providing appropriate inputs, you can test the behavior of service methods and validate their output.

We've already started by testing our service's `add()` method in the previous section. As repetition is educational, we'll be implementing other methods, namely subtraction, multiplication, and division.

Let's open our `calculator.component.spec.ts` file and continue writing our tests linked to our business logic, following TDD principles.

We're about to write our red test for subtracting two numbers. As we already have experience with the addition functionality, we can draw inspiration from it:

```
it('should substract two numbers correctly', () ⇒ {
    spyOn(calculatorService, 'substract').and.callThrough();
    calculator.substract(2, 3);
    expect(calculatorService.substract).toHaveBeenCalledWith(2, 3);
    expect(calculator.result).toBe(-1);
});
```

Figure 3.12 – subtract method test code without some errors

Obviously, we have red because the subtract method doesn't yet exist. Our test won't even run. Let's take a look at our calculator.service.ts service and add it. Keep in mind that we need to write as little code as possible:

```
substract(a: number, b: number): number {
    return a - b;
}
```

Figure 3.13 – subtract method declaration in calculator.service.ts

One the other hand, in our calculator.component.spec.ts, note that there's less red, but there's still some, as shown in the following figure:

```
it('should substract two numbers correctly', () ⇒ {
    spyOn(calculatorService, 'substract').and.callThrough();
    calculator.substract(2, 3);
    expect(calculatorService.substract).toHaveBeenCalledWith(2, 3);
    expect(calculator.result).toBe(-1);
});
```

Figure 3.14 – subtract method test before implementing it in calculator.component.ts

Our calculator.component.ts component is missing the substract() method. Just as we did with add(), we're going to take inspiration from it:

```
substract(a: number, b: number): void {
    this.result = this.calculatorService.substract(a, b);
}
```

Figure 3.15 – Adding the substract method in calculator.component.ts

The result in our `calculator.component.spec.ts` test file is as expected:

```
it('should substract two numbers correctly', () ⇒ {
  spyOn(calculatorService, 'substract').and.callThrough();
  calculator.substract(2, 3);
  expect(calculatorService.substract).toHaveBeenCalledWith(2, 3);
  expect(calculator.result).toBe(-1);
});
```

Figure 3.16 – subtract method test case without error

When we take a tour of the terminal, we get this preview:

```
✓ Browser application bundle generation complete.
Chrome 115.0.0.0 (Mac OS 10.15.7): Executed 8 of 8 SUCCESS (0.11 secs / 0.097 se
cs)
TOTAL: 8 SUCCESS

=============================== Coverage summary ===============================
Statements   : 100% ( 11/11 )
Branches     : 100% ( 0/0 )
Functions    : 100% ( 8/8 )
Lines        : 100% ( 10/10 )
================================================================================
```

Figure 3.17 – Test coverage of our app

We'll do the same exercise for multiplication and division. In our `calculator.component.spec.ts`, we'll get this:

```
it('should multiply two numbers correctly', () ⇒ {
  spyOn(calculatorService, 'multiply').and.callThrough();
  calculator.multiply(2, 3);
  expect(calculatorService.multiply).toHaveBeenCalledWith(2, 3);
  expect(calculator.result).toBe(6);
});

it('should divide two numbers correctly', () ⇒ {
  spyOn(calculatorService, 'divide').and.callThrough();
  calculator.divide(4, 2);
  expect(calculatorService.divide).toHaveBeenCalledWith(4, 2);
  expect(calculator.result).toBe(2);
});
```

Figure 3.18 – Adding multiply and divide methods test cases

Then in our `calculator.service.ts` service, we have the following:

```
multiply(a: number, b: number): number {
  return a * b;
}

divide(a: number, b: number): number {
  return a / b;
}
```

Figure 3.19 – Adding multiply and divide methods in calculator.service.ts

And finally, in our `calculator.component.ts`, we have the following:

```
multiply(a: number, b: number): void {
  this.result = this.calculatorService.multiply(a, b);
}

divide(a: number, b: number): void {
  this.result = this.calculatorService.divide(a, b);
}
```

Figure 3.20 – Adding multiply and divide methods in calculator.component.ts

In our terminal, we can see the following:

```
✔ Browser application bundle generation complete.
Chrome 115.0.0.0 (Mac OS 10.15.7): Executed 1 of 10 SUCCESS (0 secs / 0.018 secs
Chrome 115.0.0.0 (Mac OS 10.15.7): Executed 2 of 10 SUCCESS (0 secs / 0.022 secs
Chrome 115.0.0.0 (Mac OS 10.15.7): Executed 3 of 10 SUCCESS (0 secs / 0.024 secs
Chrome 115.0.0.0 (Mac OS 10.15.7): Executed 4 of 10 SUCCESS (0 secs / 0.026 secs
Chrome 115.0.0.0 (Mac OS 10.15.7): Executed 5 of 10 SUCCESS (0 secs / 0.028 secs
Chrome 115.0.0.0 (Mac OS 10.15.7): Executed 7 of 10 SUCCESS (0 secs / 0.032 secs
Chrome 115.0.0.0 (Mac OS 10.15.7): Executed 8 of 10 SUCCESS (0 secs / 0.074 secs
Chrome 115.0.0.0 (Mac OS 10.15.7): Executed 9 of 10 SUCCESS (0 secs / 0.079 secs
Chrome 115.0.0.0 (Mac OS 10.15.7): Executed 10 of 10 SUCCESS (0 secs / 0.083 sec
Chrome 115.0.0.0 (Mac OS 10.15.7): Executed 10 of 10 SUCCESS (0.099 secs / 0.083
secs)
TOTAL: 10 SUCCESS

=============================== Coverage summary ===============================
Statements    : 100% ( 15/15 )
Branches      : 100% ( 0/0 )
Functions     : 100% ( 12/12 )
Lines         : 100% ( 14/14 )
================================================================================
```

Figure 3.21 – Test coverage of our app

And in our browser, we can see the following:

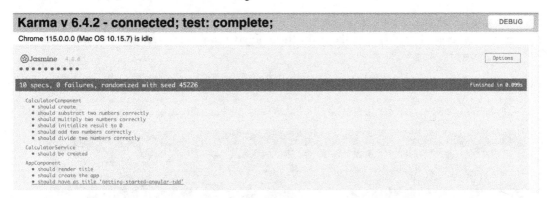

Figure 3.22 – calculator.component.ts test cases succeeded

In the next section, we'll make things a little more interesting. We're going to turn our `result` variable into an observable. This way, we won't have to call it up every time in our component's calculation methods. This will also enable us to see how to test an observable.

Using rigorous directive testing to ensure proper rendering and functionality

Angular directives play a vital role in structuring and enhancing the functionality of web applications. They allow developers to manipulate the DOM, create reusable components, and provide dynamic behavior. Directive testing is the process of verifying that directives render correctly and function as expected. By thoroughly testing directives, developers can identify and fix issues before they impact the application's overall performance and user experience.

In our current calculator application development project, we're going to use a directive to apply a color to the calculation result displayed on the screen.

Implementing color change directives

To handle color changes in our Angular calculator application, we're going to create a custom directive. Directives allow us to extend the functionality of HTML elements and encapsulate specific behaviors.

In this case, we'll create a directive called `colorChange` that will be responsible for handling color transitions. The directive will accept an input parameter specifying the color to be changed. It will then apply the CSS styles required to achieve the desired effect.

To create the directive, follow these steps:

1. Create a `directives` folder in the `core` folder of our project. So, we'll basically have `src/core/directives`, and we'll execute the following command from our terminal while in the `directives` folder:

    ```
    ng g directive color-change --skip-import
    ```

2. Then, in the `calculator.module.ts` file, we'll import our directive into the declarations table:

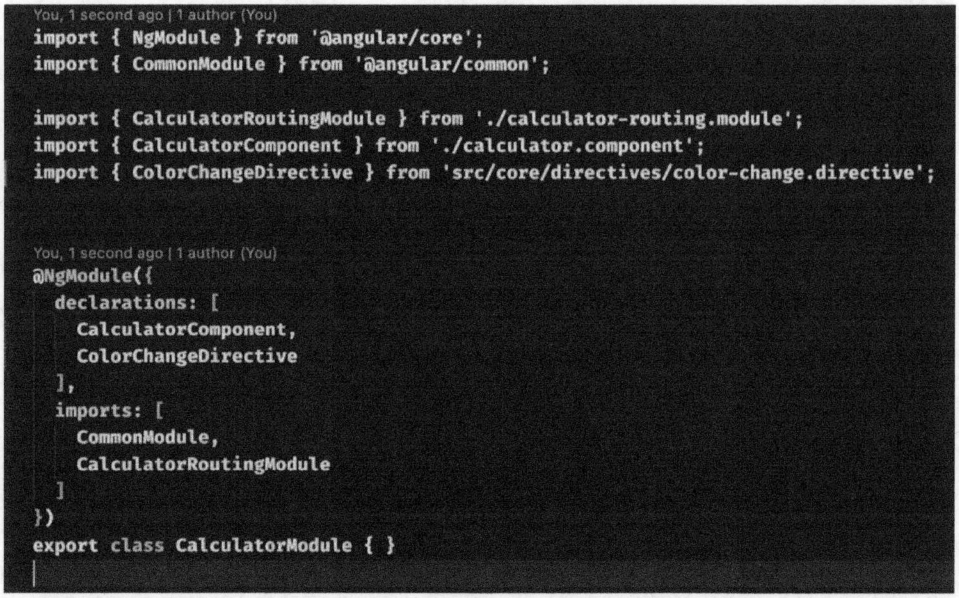

```
You, 1 second ago | 1 author (You)
import { NgModule } from '@angular/core';
import { CommonModule } from '@angular/common';

import { CalculatorRoutingModule } from './calculator-routing.module';
import { CalculatorComponent } from './calculator.component';
import { ColorChangeDirective } from 'src/core/directives/color-change.directive';


You, 1 second ago | 1 author (You)
@NgModule({
  declarations: [
    CalculatorComponent,
    ColorChangeDirective
  ],
  imports: [
    CommonModule,
    CalculatorRoutingModule
  ]
})
export class CalculatorModule { }
```

Figure 3.23 – Adding ColorChangeDirective in the declarations array of CalculatorModule

3. Then, in our `color-change.directive.ts` file, in the `selector` attribute, we'll replace `appColorChange` with a simple `colorChange`:

```
import { Directive } from '@angular/core';

@Directive({
  selector: '[appColorChange]'
})
export class ColorChangeDirective {

  constructor() { }

}
```

```
import { Directive } from '@angular/core';

@Directive({
  selector: '[colorChange]'
})
export class ColorChangeDirective {

  constructor() { }

}
```

Before   **After**

Figure 3.24 – Changing the selector name of ColorChangeDirective

Now that we have created the `colorChange` directive, let's move on to the next section, where will write tests for this newly created directive.

Writing tests for the colorChange directive

As we are following the TDD approach, our tests should check the use of our directive.

When we apply a color to our HTML content through the directive, that color should change. Our test will naturally fail because we haven't yet written the appropriate code for our `colorChange` directive.

Subsequently, we'll write the minimum amount of code necessary for our test to pass, and refactor it if necessary.

In our `color-chnage.directive.spec.ts` file, we have the following:

```
import { ColorChangeDirective } from './color-change.directive';

describe('ColorChangeDirective', () => {
  it('should create an instance', () => {
    const directive = new ColorChangeDirective();
    expect(directive).toBeTruthy();
  });
});
```

We're going to modify the previous code and complete the test suites in line with the expected logic.

In the previous code, when our directive is created, we notice that the written test checks whether the instance exists by creating an object. In our scenario, we won't do this. We'll declare our directive directly in the `configureTestingModule` method, which guarantees its existence and the possibility of accessing it without passing through the constructor. This will give us the following result:

```
import { TestBed } from '@angular/core/testing';
import { ColorChangeDirective } from './color-change.directive';

describe('ColorChangeDirective', () => {
  beforeEach(async () => {
    await TestBed.configureTestingModule({
      declarations: [ColorChangeDirective],
    }).compileComponents();
  });
});
```

The preceding code will be our starting point. Now for a quick reminder. When we want to use a directive on an HTML tag that takes a property as a parameter, here's what it looks like:

```
<p [colorChange]="color"> </p>
```

According to the preceding code, colorChange is our directive. It takes color as a parameter. This implies that color is an attribute of our component. So, we're going to call our CalculatorComponent for our test suite, linked to the directive, so that we can interact with it. Here's what it looks like:

```
import { ComponentFixture, TestBed } from '@angular/core/testing';
import { ColorChangeDirective } from './color-change.directive';
import { CalculatorComponent } from 'src/app/calculator/calculator.
component';

describe('ColorChangeDirective', () => {
  let fixture: ComponentFixture<CalculatorComponent>;
  let calculator: CalculatorComponent;

  beforeEach(async () => {
    await TestBed.configureTestingModule({
      declarations: [ColorChangeDirective, CalculatorComponent],
    }).compileComponents();

    fixture = TestBed.createComponent(CalculatorComponent);
    calculator = fixture.componentInstance;
    fixture.detectChanges();
  });
});
```

We know that we need to select our paragraph, p, in the component CalculatorComponent to change the color of our paragraph, p, as we see fit. As we only have one paragraph in the component, here's how we can proceed:

```
import { ComponentFixture, TestBed } from '@angular/core/testing';
import { By } from '@angular/platform-browser';
import { ColorChangeDirective } from './color-change.directive';
import { CalculatorComponent } from 'src/app/calculator/calculator.
component';

describe('ColorChangeDirective', () => {
  let fixture: ComponentFixture<CalculatorComponent>;
```

```
let calculator: CalculatorComponent;

beforeEach(async () => {
  await TestBed.configureTestingModule({
    declarations: [ColorChangeDirective, CalculatorComponent],
  }).compileComponents();

  fixture = TestBed.createComponent(CalculatorComponent);
  calculator = fixture.componentInstance;
  fixture.detectChanges();
});

it('should apply the specified color', () => {
  const element: HTMLElement = fixture.debugElement.query(By.
css('p')).nativeElement;
  const color: string = 'red';
  calculator.color = color;
  fixture.detectChanges();

  expect(element.style.color).toBe(color);
});
});
```

In the preceding code, we've managed to select the paragraph using By. So, since the color attribute will be used to define the color of our paragraph, we will use it in our test suite. The code editor highlights color in red because we haven't yet declared it in our component. We'll make the necessary changes to our component next.

In the CalculatorComponent class, we'll declare the color attribute:

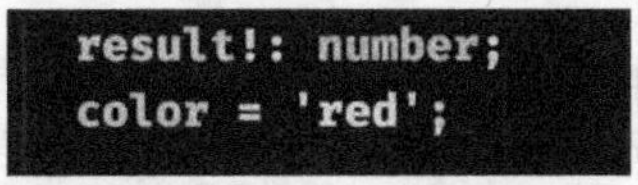

```
result!: number;
color = 'red';
```

Figure 3.25 – Adding the color attribute in calculator.component.ts

In the HTML file, we have the following:

```
<p [colorChange]="color"> {{ result }} </p>
```

In the preceding code, notice that `[colorChange]="color"` is considered an error in our HTML template:

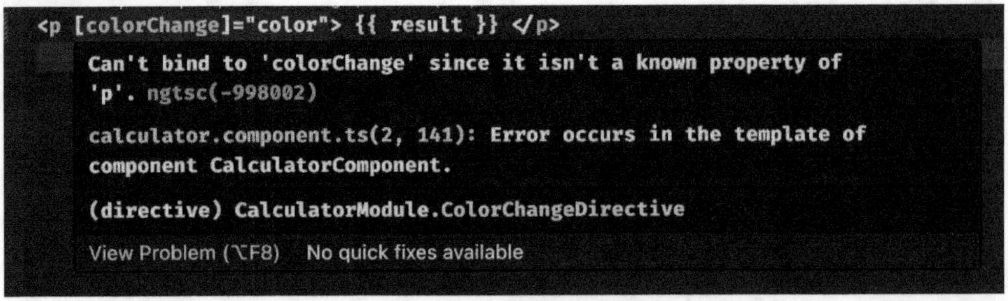

```
<p [colorChange]="color"> {{ result }} </p>
    Can't bind to 'colorChange' since it isn't a known property of
    'p'. ngtsc(-998002)

    calculator.component.ts(2, 141): Error occurs in the template of
    component CalculatorComponent.

    (directive) CalculatorModule.ColorChangeDirective

View Problem (⌥F8)    No quick fixes available
```

Figure 3.26 – Adding a colorChange directive with an error

This is normal, as our directive is missing a declaration. Since the direction takes an attribute as a parameter, we need to declare it.

Here's what we need to do in our `color-change.directive.ts` directive:

```
import { Directive, Input } from '@angular/core';

@Directive({
  selector: '[colorChange]',
})
export class ColorChangeDirective {
  @Input() colorChange!: string;
  constructor() {}
}
```

In our HTML template for the `calculator` component, there are no more errors:

```
<p [colorChange]="color"> {{ result }} </p>
```

Figure 3.27 – Adding a colorChange directive without errors

Nevertheless, we'll always have `calculator.component.spec.ts` test cases that have failed, and we'll display this on the screen when we run our tests:

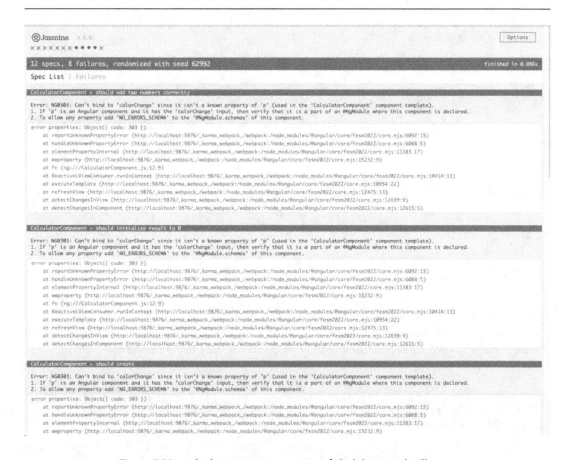

Figure 3.28 – calculator.component.ts test failed due to colorChange

To fix this, we need to declare our directive in the test file of our `CalculatorComponent`, namely `calculator.component.spec.ts`:

```
beforeEach(async () => {
  await TestBed.configureTestingModule({
    declarations: [CalculatorComponent, ColorChangeDirective],
    providers: [CalculatorService]
  }).compileComponents();
```

Figure 3.29 – Updating the beforeEach method in calculator.component.spec.ts

We'll then have just one error:

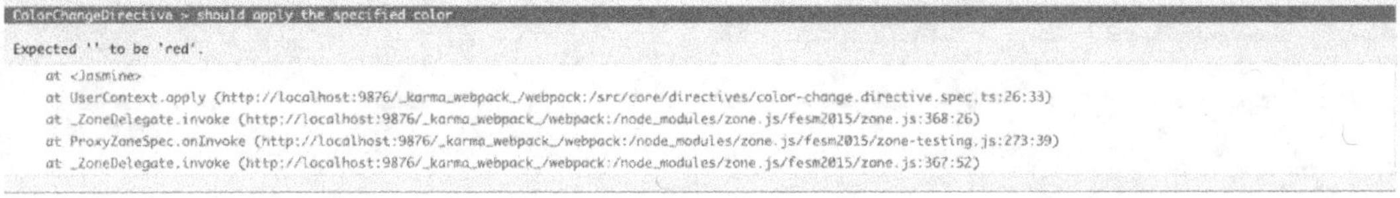

```
ColorChangeDirective > should apply the specified color
Expected '' to be 'red'.
    at <Jasmine>
    at UserContext.apply (http://localhost:9876/_karma_webpack_/webpack:/src/core/directives/color-change.directive.spec.ts:26:33)
    at _ZoneDelegate.invoke (http://localhost:9876/_karma_webpack_/webpack:/node_modules/zone.js/fesm2015/zone.js:368:26)
    at ProxyZoneSpec.onInvoke (http://localhost:9876/_karma_webpack_/webpack:/node_modules/zone.js/fesm2015/zone-testing.js:273:39)
    at _ZoneDelegate.invoke (http://localhost:9876/_karma_webpack_/webpack:/node_modules/zone.js/fesm2015/zone.js:367:52)
```

Figure 3.30 – calculator.component.ts test failed due to colorChange

This error is due to the fact that we haven't yet written the logic for our directive. We need to write the minimum amount of code for the test to pass:

```
import { Directive, ElementRef, Input, OnInit, Renderer2 } from '@
angular/core';

@Directive({
  selector: '[colorChange]',
})
export class ColorChangeDirective implements OnInit {
  @Input() colorChange!: string;
  constructor(private elementRef: ElementRef, private renderer:
Renderer2) {}

  ngOnInit() {
    this.renderer.setStyle(this.elementRef.nativeElement, 'color',
this.colorChange);
  }
}
```

In the preceding code, we implemented the ngOnInit() lifecycle to ensure that the directive is loaded into the DOM. Then we injected dependencies into the constructor, namely ElementRef and Renderer2, to manipulate the HTML element and apply the style to it. The result can be seen in the following screenshot:

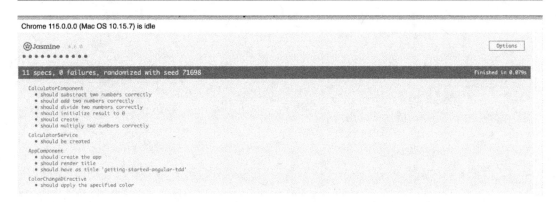

Figure 3.31 – ColorChangeDirective test succeeded

By implementing custom directives and writing rigorous tests following the TDD principles, we can ensure that our application behaves as expected, providing users with a visually appealing and interactive experience.

Summary

This chapter covered various aspects of component testing, including initializing components, rendering templates, handling events, and manipulating the DOM. It explained how to use Angular's testing utilities, such as `TestBed` and `ComponentFixture`, for setting up and interacting with components during testing. It provided insights into testing Angular services, including testing service methods and handling dependencies. It explored the testing of Angular directives, focusing on testing directive behavior and interactions with the DOM. It explained how to test directive attributes, inputs, and outputs effectively.

In the next chapter, we'll look at how to mock and stub dependencies in Angular.

4

Mocking and Stubbing Dependencies in Angular Tests

To write effective and reliable tests for Angular applications, it's essential to understand how to handle dependencies. Dependencies can often introduce complexity and make testing difficult. However, by taking advantage of techniques such as mocking and stubbing, we can better control our tests and ensure the accuracy and stability of our application.

In this chapter, we'll explore the concept of spies and method substitutes. Spies allow us to monitor and verify the behavior of dependencies during testing. We'll learn how to create spies using the Jasmine test framework and use them to find out whether certain methods have been called, how many times they've been called, and with what parameters. In addition, we'll discover the power of method substitutes, which allow us to replace the implementation of a method with our own custom logic.

Next, we'll take a look at `TestBed` providers and how they allow us to inject simulated dependencies into our tests. TestBed is a powerful Angular testing utility that allows us to create a test module and configure it with the necessary dependencies. We're going to learn how to create and configure TestBed providers to replace real dependencies with simulated versions. This technique allows us to isolate the component or service under test and control the behavior of its dependencies.

Finally, we'll explore how to handle asynchronous operations and complex scenarios when setting up dependencies. We'll discover the `async` and `fakeAsync` utilities provided by the Angular testing framework and see how to use them to manage asynchronous code in our tests. In addition, we'll discuss strategies for handling complex scenarios, such as dependencies with multiple methods or dependencies that require specific initialization steps.

In summary, here are the main topics that will be covered in this chapter:

- Monitoring and controlling dependency calls using method stubs and stubs
- Injecting mocked dependencies using TestBed providers
- Handling async operations and complex scenarios

Technical requirements

To follow along with the examples and exercises in this chapter, you will need to have a basic understanding of Angular and TypeScript. You will also need the following:

- Node.js and npm installed on your computer
- Angular CLI installed globally
- A code editor, such as Visual Studio Code, installed on your computer

The code files of this chapter can be found at `https://github.com/PacktPublishing/Mastering-Angular-Test-Driven-Development/tree/main/Chapter%204`.

Monitoring and controlling dependency calls using method stubs and spies

One crucial aspect of testing in Angular applications is the ability to monitor and control dependency calls. Dependencies are external resources or services that a piece of code relies on to function correctly. Monitoring and controlling these dependency calls allows developers to ensure that their code interacts correctly with external systems and handles different scenarios gracefully.

Spies and method stubs are two powerful techniques within Angular's testing framework that enable developers to achieve this level of control. Spies allow developers to monitor function calls, record information about those calls, and assert expectations about their usage. On the other hand, method stubs provide a way to replace real dependencies with simplified versions, allowing developers to control the behavior of those dependencies during testing.

By using spies, developers can verify that the correct functions are called with the right parameters and that they are called the expected number of times. This is particularly useful when testing code that interacts with external APIs or databases. Method stubs, on the other hand, enable developers to simulate different scenarios and provide predefined responses to method calls. This allows for thorough testing of edge cases and ensures the code's robustness.

In this section, we'll explore the concepts of spies and method stubs in the Angular testing framework. We'll dive into their applications and show their usefulness in monitoring and controlling dependency calls. Still based on our project related to a calculator application, we'll demonstrate how spies and method substitutes can be used to create reliable and complete tests, with an emphasis on the principles of **test-driven development** (**TDD**).

Method stubs and spies

Method stubs, also known as fake or dummy objects, are used to replace real dependencies with simplified versions during testing. By providing predefined responses to method calls, method substitutes enable developers to isolate and control the behavior of the code under test.

In the calculator application, let's consider a scenario in which the user performs a division operation with a divisor equal to zero. We want to make sure that the application handles this scenario correctly. By creating a method plug for the divide function, we can simulate the divide-by-zero scenario and check that the application displays an appropriate error message.

At present, our calculator's division operation does not handle the exception related to division by zero.

In our `calculator.component.spec.ts` test file, we're going to add the test that allows us to raise this exception. Since we're following TDD principles, the test should fail naturally.

After running our test, we notice that the test has indeed failed, as shown in the following screenshot:

```
it('should raise an exception when dividing by zero', () ⇒ {
    spyOn(calculatorService, 'divide').and.callThrough();
    expect(() ⇒ calculator.divide(10, 0)).toThrowError(
        'Cannot divide by zero'
    );
    expect(calculatorService.divide).toHaveBeenCalledWith(10, 0);
});
```

Figure 4.1 – Division-by-zero test case

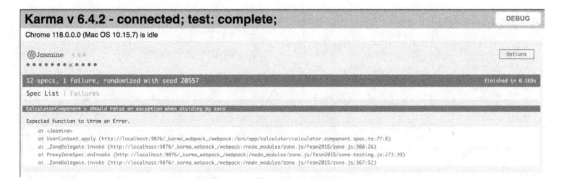

Figure 4.2 – Division-by-zero test case failed

To correct this, we need to update our `CalculatorService`. In our present scenario, there's a clever approach to refrain from direct interaction with our core service and ensure that everything is operational prior to any such action. This approach involves the utilization of a method stub concept.

Basically, we'll point our `CalculatorService` service at a fake service that will enable us to check the correctness of the logic we want to implement before modifying the service itself. In fact, this fake service will simply be a stub method that replaces the classic division of our basic `CalculatorService` service. To start with, you'll need to comment out all the tests linked to other operators in our `calculator.component.spec.ts` file, if you already have some. Next, we'll declare this stub method:

```
const calculatorServiceStub = {
  divide: (a: number, b: number) => {
      if (b === 0) {
        throw new Error('Cannot divide by zero');
      }
    return a / b;
  },
};
```

Then, in the `describe` method, within the `configureTestingModule` method, we will replace our `CalculatorService` provider with the following:

```
providers: [
        { provide: CalculatorService, useValue: calculatorServiceStub
},
      ],
```

Finally, here is the test case for our fake `calculatorServiceStub` service, which contains our stub `divide` method:

```
it('should raise an exception when dividing by zero', () => {
    spyOn(calculatorService, 'divide').and.callThrough();
    expect(() => calculator.divide(10, 0)).toThrowError(
      'Cannot divide by zero'
    );
    expect(calculatorService.divide).toHaveBeenCalledWith(10, 0);
  });
```

Here's the complete code for implementing our fake service:

```
import { ComponentFixture, TestBed } from '@angular/core/testing';

import { CalculatorComponent } from './calculator.component';
import { CalculatorService } from 'src/core/services/calculator.
service';
import { ColorChangeDirective } from 'src/core/directives/color-
change.directive';

const calculatorServiceStub = {
  divide: (a: number, b: number) => {
    if (b === 0) {
      throw new Error('Cannot divide by zero');
```

```
    }
    return a / b;
  },
};

describe('CalculatorComponent', () => {
  let calculator: CalculatorComponent;
  let fixture: ComponentFixture<CalculatorComponent>;
  let calculatorService: CalculatorService;

  beforeEach(async () => {
    await TestBed.configureTestingModule({
      declarations: [CalculatorComponent, ColorChangeDirective],
      providers: [
        { provide: CalculatorService, useValue: calculatorServiceStub
},
      ],
    }).compileComponents();

    fixture = TestBed.createComponent(CalculatorComponent);
    calculator = fixture.componentInstance;
    calculatorService = TestBed.inject(CalculatorService);
    fixture.detectChanges();
  });

  it('should create', () => {
    expect(calculator).toBeTruthy();
  });

  it('should initialize result to 0', () => {
    calculator.ngOnInit();
    expect(calculator.result).toEqual(0);
  });

  it('should divide two numbers correctly', () => {
    spyOn(calculatorService, 'divide').and.callThrough();
    calculator.divide(4, 2);
    expect(calculatorService.divide).toHaveBeenCalledWith(4, 2);
    expect(calculator.result).toBe(2);
  });

  it('should raise an exception when dividing by zero', () => {
    spyOn(calculatorService, 'divide').and.callThrough();
```

```
    expect(() => calculator.divide(10, 0)).toThrowError(
      'Cannot divide by zero'
    );
    expect(calculatorService.divide).toHaveBeenCalledWith(10, 0);
  });
});
```

Let's go through the code.

The `calculatorServiceStub` object is created to mock the `divide` method of the `CalculatorService` service. The `divide` method takes two parameters, a and b, and performs the division operation. In this case, the stub checks whether b is equal to zero. If it is, an error is thrown to simulate the division-by-zero scenario.

The last `expect` statement checks whether the `result` property of the component is equal to `'Division by zero'`. This verifies that the error message is correctly displayed when division by zero occurs.

Notice that our test execution has failed. This is a typing problem, as `result` is of the type `number` and not `string`.

So, we're going to write the minimum amount of code we need in our `CalculatorComponent` to solve the problem:

```
result!: number | string;
```

Figure 4.3 – Updating the declaration of the result property

After running the tests, notice that all the tests have turned green, as shown in our screenshot:

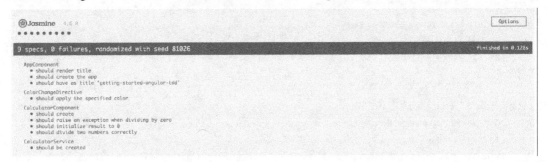

Figure 4.4 – Division-by-zero test case succeeded

Now we're going to refactor, as required by the principles of TDD. We're going to create a `stubs` folder in our project's `core` folder and create a `calculator.service.stub.ts` file in it.

Figure 4.5 – calculator.service.stub.ts file arborescence

Once the file has been created, we will move the source code of our fake stub-based service from `calculator.component.spec.ts` into our `calculator.service.stub.ts` file like this:

```
calculator.service.stub.ts U  X
1  const calculatorServiceStub = {
   Complexity is 4 Everything is cool!
2    divide: (a: number, b: number) => {
3      if (b === 0) {
4        throw new Error('Cannot divide by zero');
5      }
6      return a / b;
7    },
8  };
```

Figure 4.6 – calculator.service.stub.ts code source

Since it's a service, we'll be using the dependency injection technique, as we'll see in more detail in the next section. At this stage, here's what we'll do in a couple of steps:

1. Create a class in `calculator.service.stub.ts` called `CalculatorServiceStub`.

2. Implement all the operator methods of our calculator application.

    Here's what the source code looks like:

    ```
    export class CalculatorServiceStub {
      add(a: number, b: number): number {
        return a + b;
      }

      substract(a: number, b: number): number {
        return a - b;
    ```

```
    }

    multiply(a: number, b: number): number {
      return a * b;
    }

    divide(a: number, b: number): number | Error {
      if (b === 0) {
        throw new Error('Cannot divide by zero');
      }
      return a / b;
    }
  }
```

3.   After updating our fake service, we'll go into our `calculator.component.spec.ts` test file to replace the provider like this:

```
import { CalculatorServiceStub } from 'src/core/stubs/
calculator.service.stub';
...
providers: [
        { provide: CalculatorService, useClass:
CalculatorServiceStub },
      ],
...
```

4.   Now we can uncomment all the methods in our `calculator.component.spec.ts` file, except the `it('should display error message for division by zero')` test case. Notice that all our tests are green, as shown in the following figure:

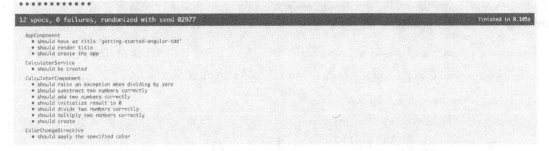

Figure 4.7 – CalculatorComponent succeeded using CalculatorServiceStub test cases

Spies and method stubs are powerful tools in Angular's testing framework that enable developers to monitor and control the behavior of dependencies in their applications. By incorporating these techniques into the TDD process, developers can write more comprehensive and reliable tests, ensuring the functionality and stability of their Angular applications. The calculator app example illustrates how spies and method stubs can be applied to simulate different scenarios and validate the behavior of the code under test. In the next section, we'll take a closer look at how to inject mocked dependencies using TestBed providers.

Injecting mocked dependencies using TestBed providers

Mocked dependencies are simplified versions of external services or resources that mimic the behavior of the actual dependencies. By injecting these mocked dependencies using TestBed providers, developers can control their behavior during testing, ensuring reliable and thorough testing without relying on external systems.

In this section, we will explore the concept of TestBed providers in Angular's testing framework. We will delve into their applications and showcase how they enable developers to inject mocked dependencies into their code. By doing so, we can create isolated and controlled testing environments, adhering to the principles of TDD.

By injecting mocked dependencies using TestBed providers, developers can focus on testing specific units of code without worrying about the complexities of the actual dependencies. This approach allows for easier debugging, improved test coverage, and better overall code quality.

Throughout this section, we will emphasize the principles of TDD and demonstrate the usage of TestBed providers in a practical example, using our calculator app. By understanding and utilizing TestBed providers effectively, developers can write reliable, maintainable, and thoroughly tested Angular applications.

Let's consider adding the ability to calculate the square root of a number to our calculator application, while still relying on our `CalculatorService` service. In this case, we'll focus on the square root functionality of the calculator application.

First, we need to create a mock service that mimics the behavior of the actual square root service. We can create a simple class that returns predefined square root values for different inputs. This class, named `MockSquareRootService`, will be created in the `mocks` folder, which we'll create at the same arborescence level as the `stubs` folder, as demonstrated in the following screenshot:

Figure 4.8 – mock-square-root.service.mock.ts file arborescence

Here's an example of a mock square root service:

```
export class MockSquareRootService {
  calculateSquareRoot(value: number): number {
    // Perform a predefined square root calculation based on the input
value
    return Math.sqrt(value);
  }
}
```

Next, we will configure the test module using TestBed providers to replace the real square root service with the simulated version in `calculator.component.spec.ts`. Here's an example of how to configure the test module:

```
import { MockSquareRootService } from './mock-square-root.service';
...
beforeEach(async () => {
    await TestBed.configureTestingModule({
        ...
      providers: [{ provide: CalculatorService, useClass:
MockSquareRootService }]
    }).compileComponents();
```

In the preceding code, we provide the `CalculatorService` service token and specify the `useClass` property with `MockSquareRootService` service. This configuration tells TestBed to use the mock service whenever the code under test requests an instance of the actual square root service.

Now, when we run our tests for the calculator app, any code that depends on the square root service will receive an instance of the mocked service. We can control the behavior of the service during testing, ensuring that the calculator app correctly performs square root calculations under different scenarios. Here's some example code for the `MockSquareRootService` service that mimics the behavior of the `CalculatorService` service by returning predefined square root values for different inputs:

```
export class MockSquareRootService {
  calculateSquareRoot(value: number): number {
    // Perform a predefined square root calculation based on the input
value
    if (value === 4) {
      return 2;
    } else if (value === 9) {
      return 3;
    } else if (value === 16) {
      return 4;
    } else {
      throw new Error('Invalid input');
```

```
        }
    }
}
```

Let's finish writing our test on the square roots of numbers by adding this sequence to our test file, `calculator.component.spec.ts`:

```
it('should calculate the square root correctly', () => {
    spyOn(calculatorService, 'squareRoot').and.callThrough();
    calculator.squareRoot(16);
    expect(calculatorService.squareRoot).toHaveBeenCalledWith(16);
    expect(calculator.result).toBe(4);
});
```

To make it functional, you need to add the `squareRoot()` method to `calculator.component.ts` and `calculator.service.ts`. I won't do this in the current project, as the aim is to show how to set up a mock briefly.

TestBed providers in Angular allow you to inject mocked dependencies into your tests. This is a powerful feature that enables you to control the behavior of external dependencies and isolate the code under test.

When configuring the testing module using `TestBed.configureTestingModule`, you can provide a list of providers that specify the tokens for the dependencies you want to mock. You can then use either the `useClass` or `useValue` property to provide a mock or stub implementation for each dependency.

By providing a mock implementation, you can define the behavior of the dependency during testing. This allows you to simulate different scenarios and control the return values or error conditions of the dependency. You can also verify the interaction between the code under test and the dependency by using spies or other testing techniques.

Injecting mocked dependencies using TestBed providers helps to improve the reliability and stability of your tests. It allows you to focus on testing the specific functionality of your code without relying on the actual implementation of external dependencies. This makes your tests more deterministic and less prone to failures caused by changes in the behavior of the dependencies.

Overall, TestBed providers provide a convenient way to inject mocked dependencies into your Angular tests. They allow you to control the behavior of external dependencies and isolate the code under test, resulting in more reliable and focused tests. We'll be taking a hands-on approach in the chapters to come. The aim here is to understand the relevance of asynchronous tasks and why we need to be careful when implementing tests about them. In the next section, we will manage asynchronous operations and complex scenarios.

Handling async operations and complex scenarios

Testing asynchronous operations and complex scenarios is a crucial part of ensuring the reliability and functionality of modern software applications. In today's software development landscape, applications often rely on asynchronous operations, such as promises and observables, to handle data fetching, processing, and user interactions. Additionally, complex scenarios, involving intricate workflows, conditional logic, and multiple dependencies, need to be thoroughly tested to ensure the application behaves as expected in various scenarios.

Testing these asynchronous operations and complex scenarios requires the use of specialized techniques and tools to handle the unique challenges they present. In the context of Angular, a popular JavaScript framework, developers have access to a comprehensive testing framework that provides powerful utilities for testing such scenarios.

In this section, we will explore the importance of testing asynchronous operations, such as promises and observables, and complex scenarios in Angular applications. We will delve into various techniques and best practices for effectively testing these scenarios, ensuring reliable and comprehensive test coverage.

Understanding asynchronous operations

Asynchronous operations are tasks that can be executed independently of the main program flow. They are typically used to handle time-consuming operations, such as network requests, file I/O, or database queries. Instead of waiting for these operations to complete, the program can continue executing other tasks, improving overall performance and responsiveness.

One common approach to handling asynchronous operations is through callbacks. A callback is a function that is passed to another function as an argument and executed once the asynchronous operation is complete. This allows us to define what should happen after the operation finishes. However, managing callbacks can lead to callback hell, making the code hard to read and maintain. To solve this problem, `promises` were born.

Promises provide a more structured way to handle asynchronous operations. A promise represents the eventual completion or failure of an asynchronous operation and allows us to attach callbacks to handle these outcomes. Promises offer a more readable and maintainable way to work with asynchronous code by chaining methods, such as `.then()` and `.catch()`.

However, a new way of implementing promises has been introduced in new versions of JavaScript. `async/await` is a clean and concise syntax for working with asynchronous operations. It allows us to write asynchronous code that looks like synchronous code, making it easier to reason about and maintain. Using the `async` keyword, we can define functions that can pause and resume execution using the `await` keyword, which waits for a promise to be resolved or rejected.

In addition, another approach to asynchronous operations is to pass observables. Observables are a powerful tool for managing data flows and asynchronous operations in reactive programming. They represent a sequence of values that can be observed over time. Observables can output multiple values asynchronously and provide a wide range of operators for transforming, filtering, and combining data streams. They are commonly used in frameworks such as Angular to handle events, HTTP requests, and other asynchronous operations with RxJS.

Asynchronous operations and observables also enable concurrency and parallelism in our code. Concurrency refers to the ability to execute multiple tasks at the same time, while parallelism refers to executing tasks simultaneously across multiple processors or threads. Asynchronous programming and observables allow us to handle multiple operations concurrently, improving performance in applications that require heavy computation or I/O.

Handling asynchronous operations

Let's consider that our calculator application that performs addition, subtraction, multiplication, and division now has a service that performs these operations and returns the result as an observable.

First, let's assume we have another calculator service called `CalculatorAsyncService`, which we'll create in the `services` folder. We need to execute this command line in the terminal once in the `services` folder:

```
$ ng g s calculator-async
```

After executing the command, here's what we should have in the `services` folder:

Figure 4.9 – Creation of the CalculatorAsyncService

As opposed to the previous service we had to create, here we'll essentially be doing asynchronous operations, in keeping with the subject we're exploring. Our service's methods will be based on the same principle, that is, receiving the two operands related to our calculation as parameters, then performing them as observables (which emphasizes the asynchronous aspect) and returning the result at the end. Based on the principles of TDD, we'll look at what to expect using the `add()` method as an example. In our `calculator-async.service.spec.ts` test file, we'll add this test case:

```
it('should add two numbers', fakeAsync(() => {
  let result = 0;
  service.add(1, 2).subscribe((val) => {
    result = val;
```

```
    });
    expect(result).toBe(3);
  }));
```

The preceding code snippet is a unit test for a service method that adds two numbers using Angular's `fakeAsync` utility to handle asynchronous operations synchronously. Here is the code breakdown:

- `fakeAsync`: This is an Angular utility function that lets you write tests that rely on asynchronous operations synchronously. It is useful for testing code that uses observables, promises, or other asynchronous operations.

- `service.add(1, 2).subscribe((val) => { result = val ; });`: This line calls a service's `add` method, passing two numbers (1 and 2). The `add()` method is supposed to return an observable that outputs the result of the addition. The `subscribe` method is used to subscribe to this observable and manage the emitted value. In this case, the output value is assigned to the result variable.

Now we can run our favorite `ng test` command from our project terminal:

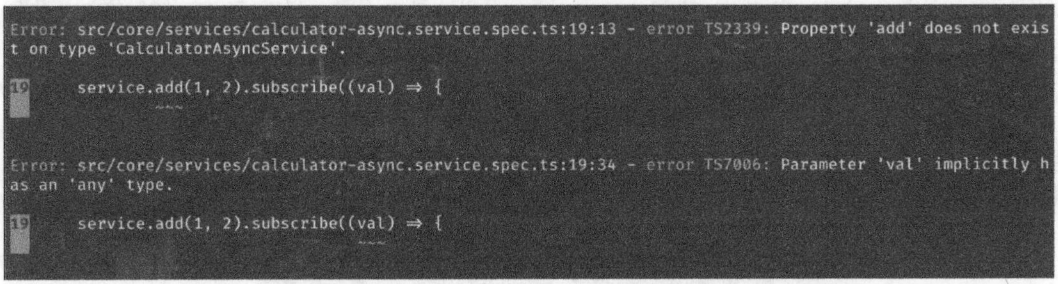

```
Error: src/core/services/calculator-async.service.spec.ts:19:13 - error TS2339: Property 'add' does not exis
t on type 'CalculatorAsyncService'.

19      service.add(1, 2).subscribe((val) => {


Error: src/core/services/calculator-async.service.spec.ts:19:34 - error TS7006: Parameter 'val' implicitly h
as an 'any' type.

19      service.add(1, 2).subscribe((val) => {
```

Figure 4.10 – Error in our test case related to the add method on the terminal

As we can see, our test failed. This is normal, as this is the red phase. We have two errors:

- The non-existence of the `add()` method in our `CalculatorAsyncService`

- The absence of the type of our `val` variable

To fix this, here's what we'll do. First, our `val` variable is the result of our calculation. It is therefore of the type number. So, we'll do the following:

```
  it('should add two numbers', fakeAsync(() => {
    let result = 0;
    service.add(1, 2).subscribe((val: number) => {
      result = val;
    });
```

```
    expect(result).toBe(3);
  }));
```

Here's how it looks on the terminal:

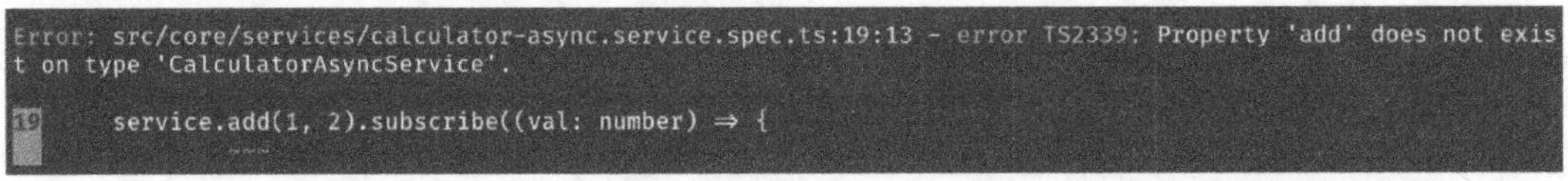

```
Error: src/core/services/calculator-async.service.spec.ts:19:13 - error TS2339: Property 'add' does not exis
t on type 'CalculatorAsyncService'.

19     service.add(1, 2).subscribe((val: number) => {
```

Figure 4.11 – Error in our test case related to the add method on the terminal

The error related to the `val` variable has now disappeared, but the error related to our service's `add()` method remains. This is normal because our `CalculatorAsyncService` doesn't yet have an `add()` method. Now we're going to write the minimum code required for our test to pass. As a reminder, our `add()` method must return an observable. Here's the code for the `add()` method to be added to our `CalculatorAsyncService`:

```
add(a: number, b: number): Observable<number> {
  return of(a + b);
}
```

Here is a breakdown of the preceding code:

- `add(a : number, b : number) : Observable<number>`: This is the method signature. The method is called `add()` and takes two parameters, a and b, both of which are numbers. The method returns an observable, which outputs a number. This indicates that the method is asynchronous and will produce a value at some point in the future.

- `return of(a + b);`: This line uses RxJS's of function to create an observable that outputs a single value and terminates. The value emitted is the result of adding a and b. The of function is a utility function that converts the given arguments into an observable sequence. In this case, it is used to create an observable that outputs the sum of a and b. Don't forget to import it.

After implementing the `add()` method, here's the result in our terminal:

```
✓ Browser application bundle generation complete.
```

Figure 4.12 – add asynchronous method test case succeeded on the terminal

And in our browser running on Karma, we have this:

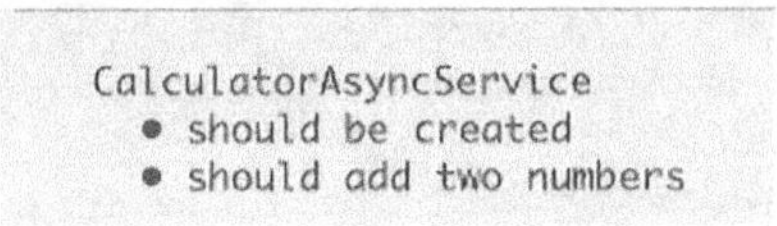

```
CalculatorAsyncService
  • should be created
  • should add two numbers
```

Figure 4.13 – add asynchronous method test case succeeded in the browser

Well done! We've just implemented our calculator's first asynchronous method using TDD principles. We'll now do the same with subtraction, multiplication, and division.

In our `calculator-async.service.spec.ts` test file, we're going to write the expected tests related to our subtraction, multiplication, and division operators. Usually, we'd do these separately, but as we've already seen with the `add()` method, we'll do them all at once. Here's what we'll get:

```
it('should subtract two numbers', fakeAsync(() => {
    let result = 0;
    service.subtract(5, 3).subscribe((val: number) => {
      result = val;
    });

    expect(result).toBe(2);
}));

it('should multiply two numbers', fakeAsync(() => {
    let result = 0;
    service.multiply(3, 4).subscribe((val: number) => {
      result = val;
    });

    expect(result).toBe(12);
}));

it('should divide two numbers', fakeAsync(() => {
    let result = 0;
    service.divide(10, 2).subscribe((val: number) => {
      result = val;
    });

    expect(result).toBe(5);
}));
```

As you may have noticed in your terminal, we have these errors:

```
Error: src/core/services/calculator-async.service.spec.ts:28:13 - error TS2339: Property 'subtract' does not
 exist on type 'CalculatorAsyncService'.

28     service.subtract(5, 3).subscribe((val: number) => {


Error: src/core/services/calculator-async.service.spec.ts:37:13 - error TS2339: Property 'multiply' does not
 exist on type 'CalculatorAsyncService'.

37     service.multiply(3, 4).subscribe((val: number) => {


Error: src/core/services/calculator-async.service.spec.ts:46:13 - error TS2339: Property 'divide' does not e
xist on type 'CalculatorAsyncService'.

46     service.divide(10, 2).subscribe((val: number) => {
```

Figure 4.14 – Error in our test case related to the subtract, multiply, and divide methods on the terminal

These errors are due to the absence of the `subtract`, `multiply`, and `divide` methods in our `CalculatorAsyncService`. As we had to do for the `add` method, we'll add the minimum amount of code needed to make our tests go green. In our `CalculatorAsyncService`, we'll add these methods:

```
subtract(a: number, b: number): Observable<number> {
  return of(a - b);
}


multiply(a: number, b: number): Observable<number> {
  return of(a * b);
}


divide(a: number, b: number): Observable<number> {
  return of(a / b);
}
```

On our terminal, we have the following:

```
Chrome 123.0.0.0 (Mac OS 10.15.7): Executed 17 of 17 SUCCESS (0.126 secs / 0.115 secs)
Chrome 123.0.0.0 (Mac OS 10.15.7) ERROR
  Disconnected Client disconnected from CONNECTED state (transport close)
15 04 2024 15:57:48.584:INFO [Chrome 123.0.0.0 (Mac OS 10.15.7)]: Connected on socket DykotjeT2isFJU8mAAAN w
ith id 83885819
Chrome 123.0.0.0 (Mac OS 10.15.7): Executed 17 of 17 SUCCESS (0.124 secs / 0.11 secs)
TOTAL: 17 SUCCESS
```

Figure 4.15 – subtract, multiply, and divide asynchronous methods test cases succeeded on the terminal

And in our browser, we have the following:

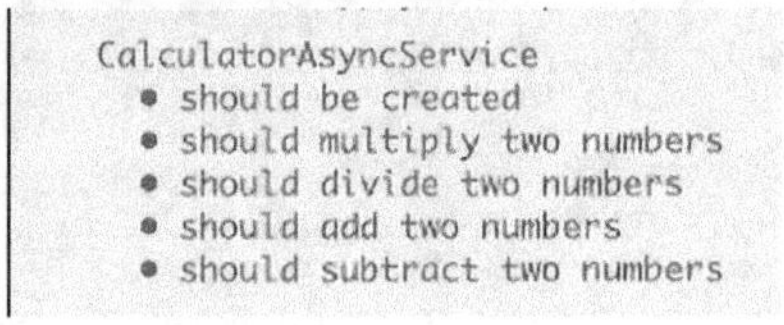

```
CalculatorAsyncService
  • should be created
  • should multiply two numbers
  • should divide two numbers
  • should add two numbers
  • should subtract two numbers
```

Figure 4.16 – subtract, multiply, and divide asynchronous methods test cases succeeded on the browser

All our tests are green!

However, there's one case we haven't yet tested at the division level. It's division by zero. As with the CalculatorService service previously, we also need to handle division by 0 by raising an exception or returning an error message. So, at the CalculatorAsyncService level, we're going to add a second test case related to division, which handles the case where we don't try to divide a number by 0:

```
it('should throw an error when dividing by zero', fakeAsync(() => {
  let error = { message: '' }; ;
  service.divide(10, 0).subscribe({
    error: (err) => (error = err),
  });

  expect(error).toBeTruthy();
  expect(error.message).toBe('Cannot divide by zero');
}));
```

Here's a breakdown of the code:

- service.divide(10, 0).subscribe({ error : (err) => (error = err), }) ;: This line calls a service's divide method with arguments 10 and 0. The divide method is expected to return an observable. The subscribe method is used to subscribe to the observable. The object passed to subscribe specifies how to handle the case of an error. If an error occurs during division (which will happen, since division by zero is not defined), the error-handling function (err) => (error = err) is executed, assigning the error message to the error variable.

- expect(error).toBeTruthy() ;: This line asserts that the error variable is true, that is, that it has a value. This is a basic check to ensure that an error has been triggered.

- expect(error.message).toBe('Cannot divide by zero') ;: This line asserts that the message property of the error object is equal to the 'Cannot divide by zero' string. This is the specific error message expected when attempting to divide by zero.

After adding our test case, let's see what happens on our terminal:

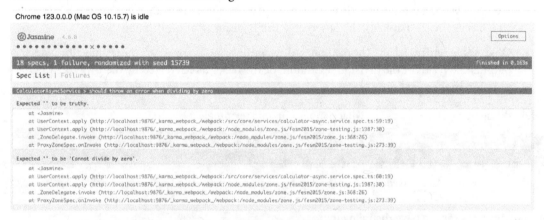

Figure 4.17 – Asynchronous division-by-0 test case failed on the terminal

And in our browser, we have the following:

Figure 4.18 – Asynchronous division-by-0 test case failed in the browser

This error is to be expected because we haven't yet handled it in our `CalculatorAsyncService` service. We're now going to write the minimal code needed to resolve this error. In our `CalculatorAsyncService` service, we're going to modify our `divide` method:

```
divide(a: number, b: number): Observable<number> {
    if (b === 0) {
        return throwError(() => new Error('Cannot divide by zero'));
    }
    return of(a / b).pipe(
        catchError((error) => {
            return throwError(() => error);
        })
    );
}
```

Here's a breakdown of the code:

- **Error handling for division by zero**: The method first checks whether b is equal to 0 using if (b === 0). If b is equal to 0, it returns an observable that immediately throws an error with the message Cannot divide by zero. To do this, we use RxJS's throwError function, which creates an observable that emits no elements and immediately issues an error notification.

- **Execute division**: If b is not 0, the method performs the a / b division operation and wraps the result in an observable using the RxJS of function. This function creates an observable that outputs the specified value, then terminates.

- **Handling other errors**: The divide operation is then routed through the catchError operator. This operator catches any errors that occur during the execution of the observable chain and allows you to handle them. In this case, the catchError operator is used to catch any errors that may occur during the divide operation and throw them back using the throwError function. This ensures that if an error occurs (other than division by zero, which is explicitly handled), the observable will issue an error notification.

- **Observable return**: Finally, the method returns the observable created by the of function, which outputs the result of the division operation, or the observable created by the throwError function in the event of an error.

After adding our test case, let's see what happens on our terminal:

```
✓ Browser application bundle generation complete.
Chrome 123.0.0.0 (Mac OS 10.15.7): Executed 18 of 18 SUCCESS (0.162 secs / 0.145 secs)
TOTAL: 18 SUCCESS
```

Figure 4.19 – Asynchronous division-by-0 test case succeeded on the terminal

And in our browser, we have the following:

```
CalculatorAsyncService
  • should divide two numbers
  • should throw an error when dividing by zero
  • should multiply two numbers
  • should be created
  • should add two numbers
  • should subtract two numbers
```

Figure 4.20 – Asynchronous division-by-0 test case succeeded on the browser

Error handling is an essential part of working with asynchronous operations. In our calculator application, if an error occurs while retrieving data or performing calculations, we can display an error message to the user and provide options for retrying or handling the error in an elegant way. The RxJS module provides error-handling mechanisms, such as the use of the catchError operator.

Emphasizing the importance of testing async operations

In a calculator app, async operations can include fetching data from an API, performing calculations asynchronously, or handling user input events. Properly testing these async operations is essential to ensure that the app functions correctly, provides accurate results, and handles errors gracefully.

Unit testing is a fundamental approach to testing individual components or functions in isolation. In the context of async operations in Angular, unit tests play a crucial role in verifying the behavior of code that handles async tasks. For example, you can write unit tests to verify that an API service correctly fetches exchange rates or that a calculation service accurately performs calculations asynchronously.

To effectively test async operations, it is essential to mock dependencies, such as API services or calculation functions. By mocking these dependencies, you can control the behavior of external services or functions during testing, allowing you to focus on the specific code that handles async operations. Angular provides tools such as TestBed and Jasmine spies to mock dependencies effectively.

Testing async operations often involves dealing with timing issues. For example, when testing a function that performs an async calculation, you need to ensure that the test waits for the calculation to be completed before making assertions. Angular provides utilities, such as `fakeAsync`, that allow you to control the timing of async operations in your tests, making it easier to write accurate and deterministic tests.

While unit testing is essential, it is equally important to perform integration testing to validate the interaction between different components in your calculator app. Integration tests can verify that async operations, such as fetching data and performing calculations, are correctly integrated into the overall functionality of the app. For example, you can write integration tests to ensure that the UI is updated correctly when async operations are complete.

Testing error handling is crucial in async operations. For example, when fetching data from an API, you need to test scenarios where the API returns an error response. By simulating error conditions in your tests, you can verify that the app handles errors gracefully, displays appropriate error messages, and provides fallback mechanisms. Angular's `HttpClient` module provides mechanisms for mocking API responses and testing different error scenarios.

End-to-end (E2E) testing is essential to validate the entire system's behavior, including the async operations in your calculator app. E2E tests simulate real-world user interactions and validate the app's functionality from a user's perspective. By writing E2E tests that cover scenarios involving async operations, you can ensure that the app functions correctly and provides a seamless user experience.

Summary

In this chapter, we covered three important topics related to testing in Angular: method stubs and spies, TestBed providers, and handling async operations and complex scenarios.

Firstly, we explored the concept of method stubs and spies, which allowed us to monitor and control the calls to dependencies in our tests. We learned how to create method stubs using Jasmine's `spyOn` function, which enabled us to replace a method's implementation with our own custom behavior. This allowed us to test our code in isolation and ensure that it behaved as expected.

Next, we delved into TestBed providers, which are used to inject mocked dependencies into our tests. We learned how to use the `TestBed.configureTestingModule` method to configure our test module and provide mocked instances of dependencies. This technique allowed us to control the behavior of dependencies and focus on testing specific scenarios without relying on real implementations.

Lastly, we tackled the challenges of handling async operations and complex scenarios in our tests. We explored techniques such as using the fakeAsync function to handle asynchronous code. These techniques enabled us to write reliable tests for scenarios involving asynchronous operations and complex dependencies.

In the next chapter, we'll learn how to test Angular's pipes, forms, and reactive programming.

Testing Angular Pipes, Forms, and Reactive Programming

One of the main features of the Angular framework is its ability to easily handle data manipulation and form entry through the use of pipes, forms, and reactive programming.

Pipes in Angular allow data to be transformed before displaying it to the user. Pipes can be used to format dates and currency, or even to apply custom logic to manipulate the data in any desired way. Testing these pipes ensures that they are functioning correctly and producing the expected output.

Forms in Angular are an essential component for collecting and validating user input. Testing Angular forms involves verifying that the form fields are correctly bound to the model, validating the input, and handling form submissions. By testing forms, developers can ensure that the form behaves as expected and provides a seamless user experience.

Reactive programming is a paradigm that handles asynchronous data streams and events. In Angular, reactive programming is commonly used with the ReactiveForms module, which provides a way to manage form state and data flow in a reactive manner.

In this chapter, we will explore the different approaches and best practices for testing Angular pipes, forms, and reactive programming. We will cover the tools and techniques available for writing effective tests, as well as common pitfalls and tips for troubleshooting. By the end of this chapter, you will have a solid understanding of how to test these crucial aspects of your Angular applications and ensure their reliability and accuracy.

In summary, here are the main topics that will be covered in this chapter:

- Testing an Angular pipe as used in our project
- Applying test-driven development to our reactive form

Technical requirements

To follow along with the examples and exercises in this chapter, you will need to have a basic understanding of Angular and TypeScript, as well as access to the following:

- Node.js and npm installed on your computer
- Angular CLI installed globally
- A code editor, such as Visual Studio Code, installed on your computer

The code files of this chapter can be found at `https://github.com/PacktPublishing/Mastering-Angular-Test-Driven-Development/tree/main/Chapter%205`.

Testing an Angular pipe as used in our project

Our calculator application is a simple tool that performs basic arithmetic operations, including addition, subtraction, multiplication, and division. Let's suppose we have a feature that allows us to represent a number as a percentage. To do this, we will be using a pipe. We'll call our pipe `percent`.

In our `core` folder, we'll create a `pipes` folder. We'll access this folder via our terminal and then run the following command to create our `percent` pipe:

```
$ ng g pipe percent -skip-import
```

After pipe generation, this is the tree structure you should see:

Figure 5.1 – The pipe folder

In relation to the contents of `percent.pipe.spec.ts`, this is what we have:

```
import { PercentPipe } from './percent.pipe';
describe('PercentPipe', () => {
  it('create an instance', () => {
    const pipe = new PercentPipe();
    expect(pipe).toBeTruthy();
  });
});
```

Following the principles of **test-driven development** (**TDD**), we will create a test suite that includes the following tests:

- A test that formats a positive number to a percentage string

- A test that formats a negative number to a percentage string

- A test that formats a decimal number to a percentage string

- A test that formats a non-number to a percentage string

Let's get started.

Positive number to percentage string formatting test

Our goal here is very simple. If we have a positive number, such as 123, its percentage format must be 12300%. In our percent.pipe.spec.ts file, we will add the following test suite:

```
it('should format a positive number to a percentage string', () => {
  const input = 123;
  const output = new PercentPipe().transform(input);
  expect(output).toBe('12300%');
});
```

In our terminal, after running the ng test command, we will have this:

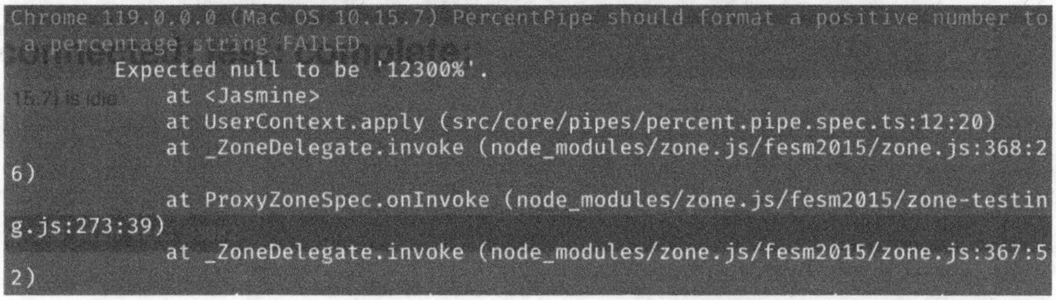

```
Chrome 119.0.0.0 (Mac OS 10.15.7) PercentPipe should format a positive number to
a percentage string FAILED
          Expected null to be '12300%'.
          at <Jasmine>
          at UserContext.apply (src/core/pipes/percent.pipe.spec.ts:12:20)
          at _ZoneDelegate.invoke (node_modules/zone.js/fesm2015/zone.js:368:2
6)
          at ProxyZoneSpec.onInvoke (node_modules/zone.js/fesm2015/zone-testin
g.js:273:39)
          at _ZoneDelegate.invoke (node_modules/zone.js/fesm2015/zone.js:367:5
2)
```

Figure 5.2 – Positive number to percentage string formatting test failed in the terminal

In our browser, we will have this result:

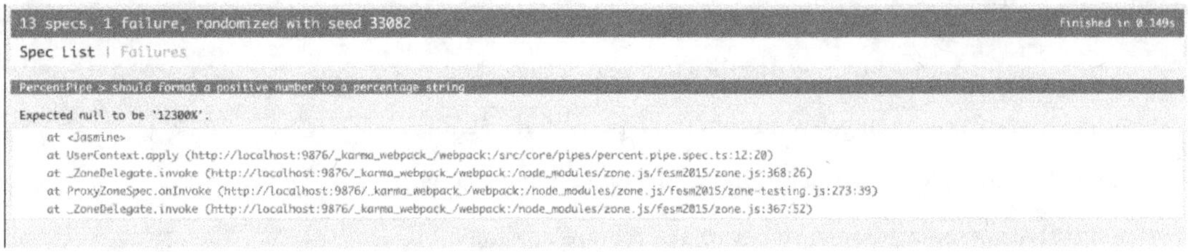

```
13 specs, 1 failure, randomized with seed 33082                                                          Finished in 0.149s
Spec List | Failures

PercentPipe > should format a positive number to a percentage string
Expected null to be '12300%'.
    at <Jasmine>
    at UserContext.apply (http://localhost:9876/_karma_webpack_/webpack:/src/core/pipes/percent.pipe.spec.ts:12:20)
    at _ZoneDelegate.invoke (http://localhost:9876/_karma_webpack_/webpack:/node_modules/zone.js/fesm2015/zone.js:368:26)
    at ProxyZoneSpec.onInvoke (http://localhost:9876/_karma_webpack_/webpack:/node_modules/zone.js/fesm2015/zone-testing.js:273:39)
    at _ZoneDelegate.invoke (http://localhost:9876/_karma_webpack_/webpack:/node_modules/zone.js/fesm2015/zone.js:367:52)
```

Figure 5.3 – Positive number to percentage string formatting test failed in the browser

It's quite normal for our test to fail. Remember, we're following the principles of TDD. So, we've thought about what's expected when our number is formatted as a percentage, and we've written the test sequence that should correspond to it. Now we're going to write the minimum amount of code required for our test to pass.

Now let's open our `percent.pipe.ts` file and find out what it contains. We'll see the following:

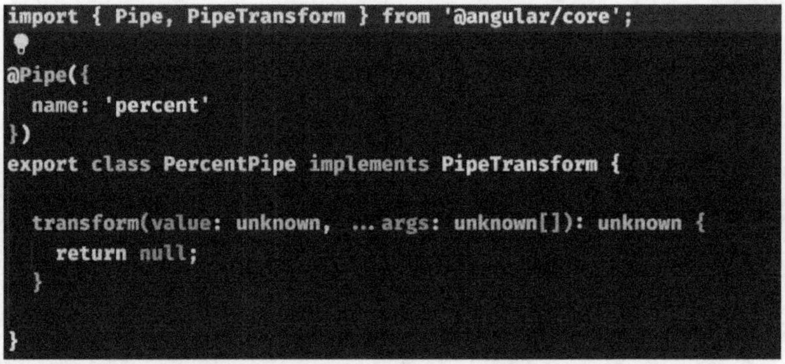

```
import { Pipe, PipeTransform } from '@angular/core';

@Pipe({
  name: 'percent'
})
export class PercentPipe implements PipeTransform {

  transform(value: unknown, ...args: unknown[]): unknown {
    return null;
  }

}
```

Figure 5.4 – Code in the percent.pipe.ts file

So, let's add our algorithm. This will allow us to take our previous test to the green stage with the minimum amount of code required:

```
transform (value: number): string {
    const formattedValue = value * 100;
    return formattedValue + '%';
  }
```

All we've done here is taken the number to be formatted as a percentage, multiplied it by 100, and concatenated it with the % symbol to get the desired rendering. Now, when we go to our test terminal, our test will turn green:

```
TOTAL: 13 SUCCESS
✓ Browser application bundle generation complete.
Chrome 119.0.0.0 (Mac OS 10.15.7): Executed 1 of 13 SUCCESS (0 secs / 0.031 secs
Chrome 119.0.0.0 (Mac OS 10.15.7): Executed 2 of 13 SUCCESS (0 secs / 0.033 secs
Chrome 119.0.0.0 (Mac OS 10.15.7): Executed 3 of 13 SUCCESS (0 secs / 0.035 secs
Chrome 119.0.0.0 (Mac OS 10.15.7): Executed 4 of 13 SUCCESS (0 secs / 0.037 secs
Chrome 119.0.0.0 (Mac OS 10.15.7): Executed 5 of 13 SUCCESS (0 secs / 0.039 secs
Chrome 119.0.0.0 (Mac OS 10.15.7): Executed 6 of 13 SUCCESS (0 secs / 0.041 secs
Chrome 119.0.0.0 (Mac OS 10.15.7): Executed 7 of 13 SUCCESS (0 secs / 0.043 secs
Chrome 119.0.0.0 (Mac OS 10.15.7): Executed 8 of 13 SUCCESS (0 secs / 0.088 secs
Chrome 119.0.0.0 (Mac OS 10.15.7): Executed 9 of 13 SUCCESS (0 secs / 0.093 secs
Chrome 119.0.0.0 (Mac OS 10.15.7): Executed 11 of 13 SUCCESS (0 secs / 0.103 sec
Chrome 119.0.0.0 (Mac OS 10.15.7): Executed 12 of 13 SUCCESS (0 secs / 0.103 sec
Chrome 119.0.0.0 (Mac OS 10.15.7): Executed 13 of 13 SUCCESS (0 secs / 0.103 sec
Chrome 119.0.0.0 (Mac OS 10.15.7): Executed 13 of 13 SUCCESS (0.121 secs / 0.103
 secs)
TOTAL: 13 SUCCESS
```

Figure 5.5 – Positive number to percentage string formatting test succeeded in the terminal

And in our browser, we will have this result:

```
PercentPipe
  ● should format a positive number to a percentage string
  ● create an instance
```

Figure 5.6 – Positive number to percentage string formatting test succeeded in the browser

In the next section, we'll look at the test for formatting a negative number as a percentage.

Negative number to percentage string formatting test

As we have seen, our objective here is also very simple. If we have a negative number, such as -123, its percentage format must be -12300%. So, in our percent.pipe.spec.ts file, we'll add the following test suite:

```
it('should format a negative number to a percentage string', () => {
  const input = -123;
  const output = new PercentPipe().transform(input);
  expect(output).toBe('-12300%');
});
```

In our terminal, after running the ng test command, we have this result:

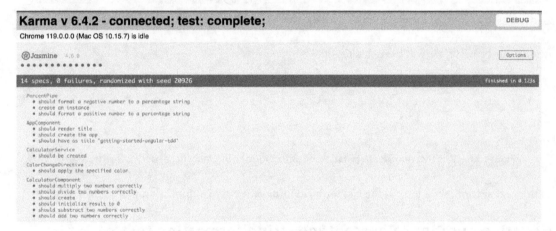

Figure 5.7 – Negative number to percentage string formatting test succeeded in the terminal

In our browser, we have the following result:

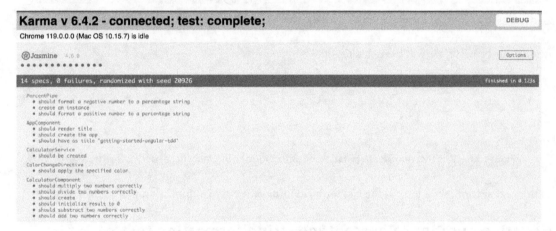

Figure 5.8 – Negative number to percentage string formatting test succeeded in the browser

We can conclude that the minimum amount of code written in our percent.pipe.ts to format positive numbers as percentages is also valid for negative numbers, as we didn't need to add any code for the test to pass.

In the next section, we'll look at the test for formatting a decimal number as a percentage.

Decimal number to percentage string formatting test

This time we'll do the same exercise with a decimal number. If we have a decimal number such as 123.45, its percentage format must be 12345%. In our percent.pipe.spec.ts file, we'll add the following test sequence:

```
it('should format a decimal number to a percentage string', () => {
  const input = 123.45;
  const output = new PercentPipe().transform(input);
```

```
    expect(output).toBe('12345%');
});
```

In our terminal, after running the ng test command, we have this:

```
→ getting-started-angular-tdd git:(main) x ng test
✓ Browser application bundle generation complete.
10 11 2023 12:11:05.468:WARN [karma]: No captured browser, open http://localhost:9876/
10 11 2023 12:11:05.519:INFO [karma-server]: Karma v6.4.2 server started at http://localhost:9876/
10 11 2023 12:11:05.519:INFO [launcher]: Launching browsers Chrome with concurrency unlimited
10 11 2023 12:11:05.524:INFO [launcher]: Starting browser Chrome
10 11 2023 12:11:14.777:INFO [Chrome 119.0.0.0 (Mac OS 10.15.7)]: Connected on socket pmO7zGPZ4GbSKnjLAAAB with id 70382
254
Chrome 119.0.0.0 (Mac OS 10.15.7): Executed 13 of 13 SUCCESS (0.177 secs / 0.145 secs)
TOTAL: 13 SUCCESS
✓ Browser application bundle generation complete.
Chrome 119.0.0.0 (Mac OS 10.15.7): Executed 14 of 14 SUCCESS (0.125 secs / 0.11 secs)
TOTAL: 14 SUCCESS
✓ Browser application bundle generation complete.
Chrome 119.0.0.0 (Mac OS 10.15.7): Executed 15 of 15 SUCCESS (0.128 secs / 0.112 secs)
TOTAL: 15 SUCCESS
```

Figure 5.9 – Decimal number to a percentage string formatting test succeeded in the terminal

In our browser, we have this result:

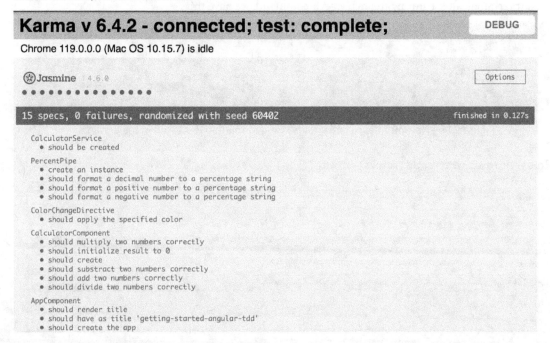

Figure 5.10 – Decimal number to a percentage string formatting test succeeded in the browser

We can conclude that the minimal amount of code written in our percent.pipe.ts to format positive and negative numbers as percentages is also valid for decimal numbers, since we didn't need to add any code for the test to pass.

In the next section, we'll look at the test for formatting a non-number expression, such as **not a number (NaN)**, as a percentage.

Non-number to percentage string formatting test

In this case, the aim is to manage an exception. Let's assume that the number to be turned into a percentage isn't a number. Normally, we'd get an error message, such as **Error**, displayed on our calculator screen. But is this the case with our current pipe? Let's take the case of a NaN and look at the result. In our `percent.pipe.spec.ts` file, we will add the following test suite:

```
it('should return an Error when the value is not a number NaN', () =>
{
    const input = NaN;
    const output = new PercentPipe().transform(input);
    expect(output).toBe('Error');
});
```

In our terminal, after running the `ng test` command, we have this:

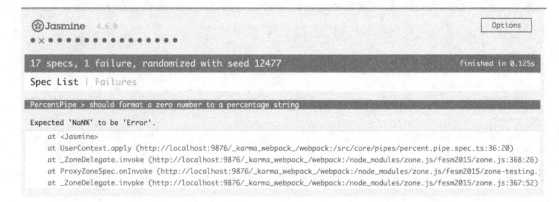

Figure 5.11 – Non-number to percentage string formatting test failed in the terminal

In our browser, we have this result:

Figure 5.12 – Non-number to percentage string formatting test succeeded in the browser

As you may have noticed, we have some errors and our test has failed. This is normal, as we haven't solved this problem using our `PercentPipe`.

To fix our test, we'll go to `percent.pipe.ts` and add the minimum code needed to handle this exception:

```
if (isNaN(value)) {
    return 'Error';
}
```

This bit of code needs to be added to the `transform()` function. The following is the complete code for the `transform()` function:

```
transform(value: number): string {
    if (isNaN(value)) {
      return 'Error';
    }

    const formattedValue = value * 100;
    return formattedValue + '%';
}
```

In our terminal, after running the `ng test` command, we will have this:

```
✔ Browser application bundle generation complete.
Chrome 119.0.0.0 (Mac OS 10.15.7): Executed 17 of 17 SUCCESS (0.124 secs / 0.105 secs)
TOTAL: 17 SUCCESS
```

Figure 5.13 – Non-number to percentage string formatting test succeeded in the terminal

In our browser, we will have this result:

```
PercentPipe
    ● should return an Error when the value is not a number NaN
    ● should format a decimal number to a percentage string
    ● create an instance
    ● should format a positive number to a percentage string
    ● should format a zero number to a percentage string
    ● should format a negative number to a percentage string
```

Figure 5.14 – Test cases for PercentPipe succeeded in the browser

The last step is to go to `CalculatorComponent` to test our `PercentPipe` in our `calculator.component.html` template. To do this, we'll declare our `PercentPipe` in the `calculator.module.ts` module so that we can use it in our template. This is what we'll get:

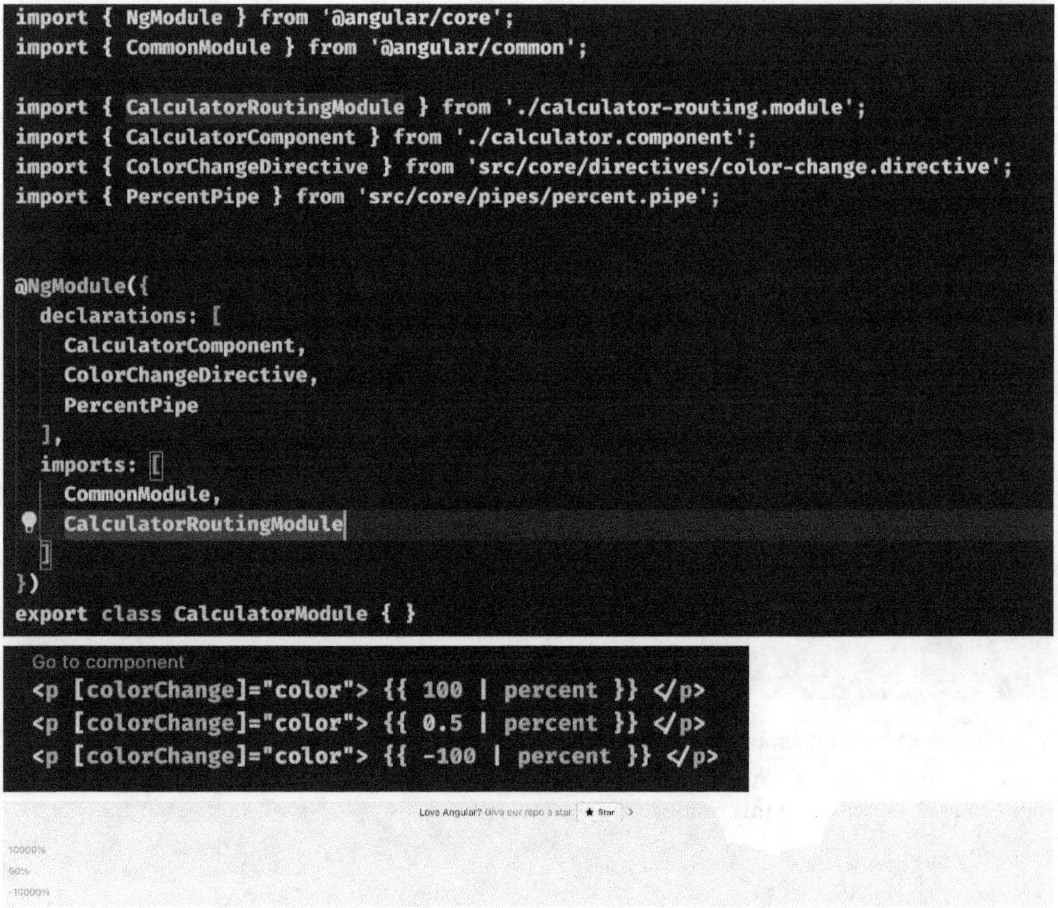

```typescript
import { NgModule } from '@angular/core';
import { CommonModule } from '@angular/common';

import { CalculatorRoutingModule } from './calculator-routing.module';
import { CalculatorComponent } from './calculator.component';
import { ColorChangeDirective } from 'src/core/directives/color-change.directive';
import { PercentPipe } from 'src/core/pipes/percent.pipe';


@NgModule({
  declarations: [
    CalculatorComponent,
    ColorChangeDirective,
    PercentPipe
  ],
  imports: [
    CommonModule,
    CalculatorRoutingModule
  ]
})
export class CalculatorModule { }
```

```html
Go to component
<p [colorChange]="color"> {{ 100 | percent }} </p>
<p [colorChange]="color"> {{ 0.5 | percent }} </p>
<p [colorChange]="color"> {{ -100 | percent }} </p>
```

Love Angular? Give our repo a star   ★ Star   >

10000%
50%
-10000%

Figure 5.15 – PercentPipe implementation in our component

Everything works perfectly!

In the next section, we'll set up our calculator's user interface. So far, there's been no user interaction. We've just been manipulating our code to play with variable states. To set up the user interface, we'll use reactive forms and see how we can apply the principles of TDD.

Implementing TDD for the Reactive Form in our calculator app

A reactive form is a type of form that uses reactive programming principles to update the form state in response to user input. Reactive forms are often used in web applications, as they can be used to create dynamic and responsive user interfaces.

In this section, we will discuss how to implement a reactive form user interface for our calculator app. We will use TDD to ensure that our form is valid and that our calculator component works as expected.

Here are the benefits of using a reactive form user interface for a calculator:

- **Dynamic and responsive**: Reactive forms can be used to create dynamic and responsive user interfaces. For example, you could use a reactive form to create a calculator that updates the result as the user enters values into the input fields.

- **Valid**: Reactive forms provide validation features that can be used to ensure that the user input is valid. For example, you could use a reactive form to create a calculator that validates that the operands are numbers, and that the operator is a valid mathematical operator.

- **Testable**: Reactive forms are easy to test using TDD. This ensures that your form is valid and that your calculator component works as expected.

To implement a reactive form user interface for using a calculator with TDD, you will need to follow the ensuing steps.

Writing a test for the calculator form

The first step is to write a test for the calculator form. This test will ensure that the form is valid when all of the fields are filled in correctly. In our `calculator.component.spec.ts`, we will add the following test suite:

```
it('should be valid when all of the fields are filled in correctly',
() => {
  const form = new FormGroup({
    operand1: new FormControl(123),
    operand2: new FormControl(456),
    operator: new FormControl('+'),
  });
  expect(form.valid).toBe(true);
});
```

Don't forget to import `ReactiveFormModule` as shown in the following code:

```
await TestBed.configureTestingModule({
    imports: [ReactiveFormsModule],
}).compileComponents();
```

As you can see, we now have a test suite linked to the management of our reactive form in the event that it is valid. When we run the `ng test` command, this will be the result:

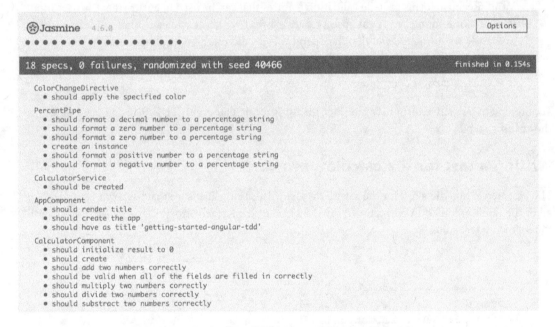

```
✓ Browser application bundle generation complete.
Chrome 119.0.0.0 (Mac OS 10.15.7): Executed 18 of 18 SUCCESS (0.155 secs / 0.14 secs)
TOTAL: 18 SUCCESS
```

Figure 5.16 – Calculator form test succeeded in the terminal

In our browser, we will have this result:

Jasmine 4.6.0   Options

18 specs, 0 failures, randomized with seed 40466   finished in 0.154s

ColorChangeDirective
 • should apply the specified color

PercentPipe
 • should format a decimal number to a percentage string
 • should format a zero number to a percentage string
 • should format a zero number to a percentage string
 • create an instance
 • should format a positive number to a percentage string
 • should format a negative number to a percentage string

CalculatorService
 • should be created

AppComponent
 • should render title
 • should create the app
 • should have as title 'getting-started-angular-tdd'

CalculatorComponent
 • should initialize result to 0
 • should create
 • should add two numbers correctly
 • should be valid when all of the fields are filled in correctly
 • should multiply two numbers correctly
 • should divide two numbers correctly
 • should substract two numbers correctly

Figure 5.17 – Calculator form test succeeded in the browser

Our test suite is on! No surprise there! In fact, we have created an instance of our form directly within our test. As we haven't interacted with our calculator component's form, we'll do that now by modifying our test suite as follows:

```
it('should be valid when all of the fields are filled in correctly',
() => {
    calculator.calculatorForm.get('operand1')?.
```

```
setValue(123);    calculator.calculatorForm.get('operand2')?.
setValue(456);
  calculator.calculatorForm.get('operator')?.setValue('+');


expect(calculator.calculatorForm.valid).toBe(true);
});
```

In our terminal, after running the ng test command, we will have this:

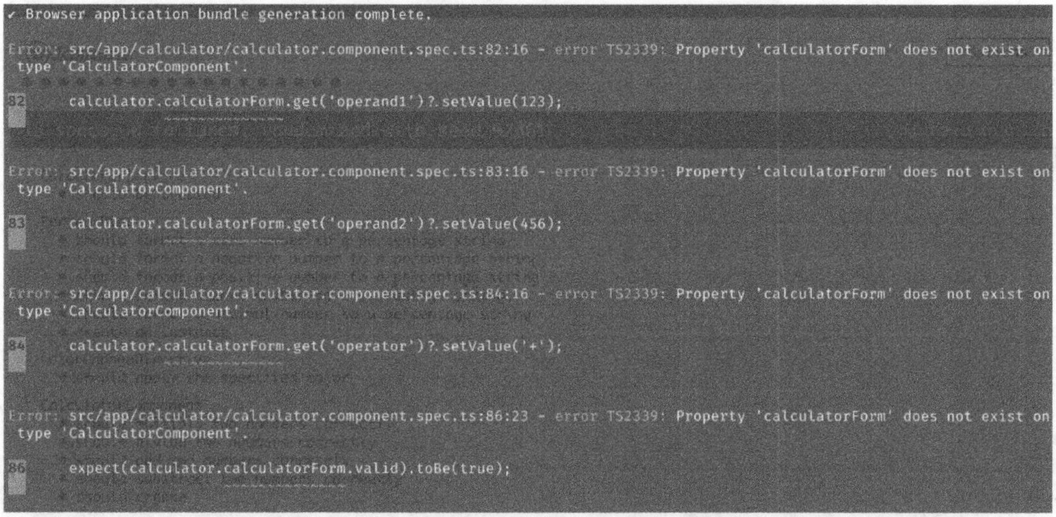

Figure 5.18 – Calculator form build failed in the terminal

The result we get is normal because calculatorForm doesn't exist in our CalculatorComponent. We need to add it as an attribute to our class as follows:

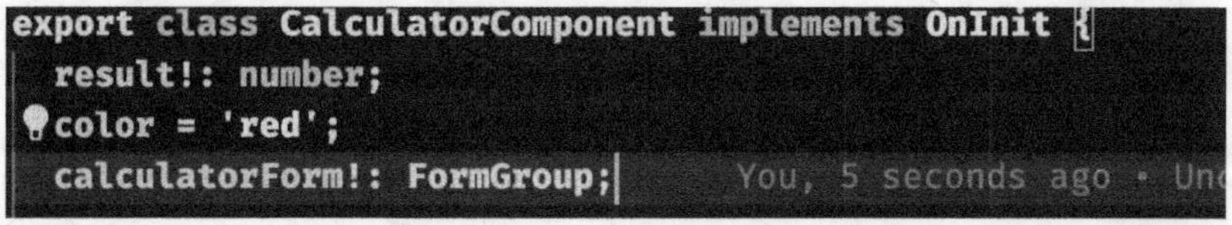

Figure 5.19 – Declaration of calculatorForm as FormGroup module in our component

Now that we have added calculatorForm, we'll also add the three input fields by initializing our calculatorForm using the following:

```
this.calculatorForm = new FormGroup({
   operand1: new FormControl(null, [Validators.required]),
   operand2: new FormControl(null, [Validators.required]),
   operator: new FormControl(null, [Validators.required]),
});
```

```
constructor(private calculatorService: CalculatorService) {
  this.calculatorForm = new FormGroup({
    operand1: new FormControl(null, [Validators.required]),
    operand2: new FormControl(null, [Validators.required]),
    operator: new FormControl(null, [Validators.required]),
  });
}
```
                                                            You, 9 second

Figure 5.20 – Initialization of our FormGroup module calculatorForm

In our terminal, after running the ng test command, we will have this:

```
✓ Browser application bundle generation complete.
Chrome 119.0.0.0 (Mac OS 10.15.7): Executed 18 of 18 SUCCESS (0.144 secs / 0.132 secs)
TOTAL: 18 SUCCESS
✓ Browser application bundle generation complete.
Chrome 119.0.0.0 (Mac OS 10.15.7): Executed 18 of 18 SUCCESS (0.108 secs / 0.091 secs)
TOTAL: 18 SUCCESS
```

Figure 5.21 – Calculator form test succeeded in the terminal

In our browser, we will have this result:

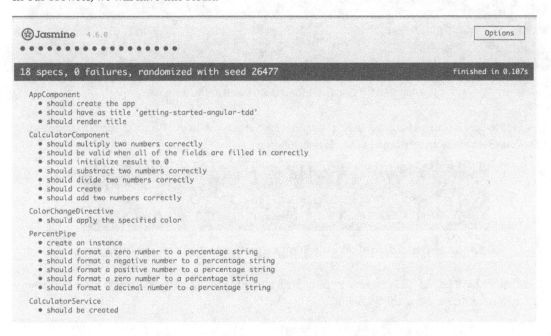

Figure 5.22 – Calculator form test succeeded in the browser

We'll do another test to make sure the opposite is true, that is, when the form is invalid.

Since all fields are mandatory, if a field is not filled in, the form must be invalid. In our `calculator.component.spec.ts` file, we'll add the following test sequence:

```
it('should be invalid when one of the field is not filled in
correctly', () => {
    calculator.calculatorForm.get('operand1')?.setValue(123);
    calculator.calculatorForm.get('operator')?.setValue('+');

    expect(calculator.calculatorForm.valid).toBe(false);
});
```

In our terminal, after running the `ng test` command, we will have this:

```
✓ Browser application bundle generation complete.
Chrome 119.0.0.0 (Mac OS 10.15.7): Executed 19 of 19 SUCCESS (0.102 secs / 0.087 secs)
TOTAL: 19 SUCCESS
```

Figure 5.23 – Calculator form test succeeded in the terminal

In our browser, we will have this result:

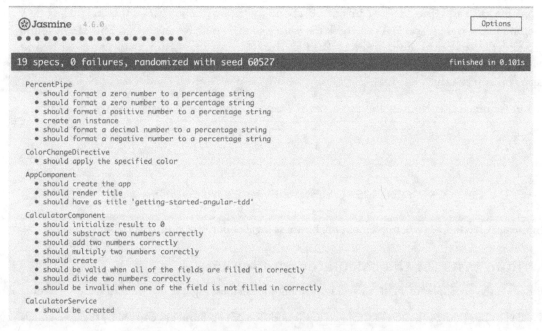

Figure 5.24 – Calculator form test succeeded in the browser

In the next section, we'll implement the user interface for our calculator application.

Implementing the user interface

Let's implement our calculator's user interface in our `calculator.component.html` file using the following:

```html
<form [formGroup]="calculatorForm">
  <input type="number" formControlName="operand1" />
  <input type="number" formControlName="operand2" />
  <select formControlName="operator">
    <option value="+">+</option>
    <option value="-">-</option>
    <option value="*">*</option>
    <option value="/">/</option>
  </select>
  <button (click)="calculate()" [disabled]="calculatorForm.invalid">
    Calculate
  </button>
  <p [colorChange]="color">{{ result | percent }}</p>
</form>
```

Don't forget to add `ReactiveFormsModule` to the `imports` array and the `calculate()` method in the right place. You can check the source code if you have any errors. This can be found at `https://github.com/PacktPublishing/Mastering-Angular-Test-Driven-Development/tree/main/Chapter%205/getting-started-angular-tdd/src/app/calculator`.

In our terminal, after running the `ng serve -o` command, we will have this in our browser:

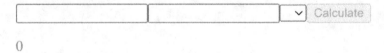

0

Figure 5.25 – Calculator user interface in the browser

In the next section, we will implement the business logic of our `calculate()` button.

Writing tests for the calculator component

Based on the principles of TDD, we will implement the business logic of our `calculate()` button.

We'll start by adding functionality related to the addition of two numbers entered by the user. In our `calculator.component.spec.ts` file, we will add the following test suite:

```typescript
it('should be added when the + operator is selected and the calculate
button is clicked', () => {
    calculator.calculatorForm.get('operand1')?.setValue(2);
    calculator.calculatorForm.get('operand1')?.setValue(3);
```

```
        calculator.calculatorForm.get('operator')?.setValue('+');
        calculator.calculate();
        expect(calculator.result).toBe(5);
    });
```

In our terminal, after running the `ng test` command, we will have this:

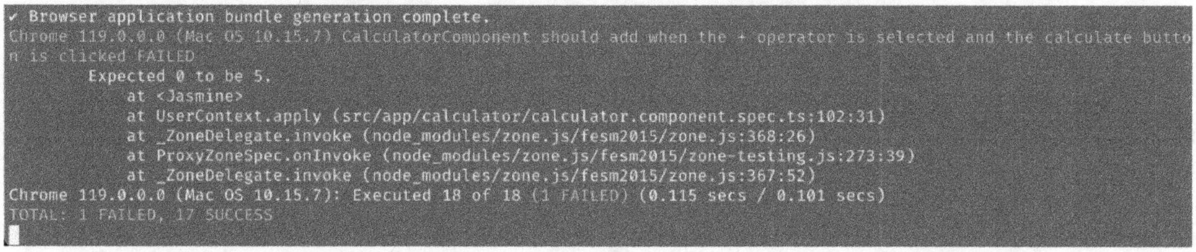

Figure 5.26 – Calculator operator (+) choice test failed in the terminal

In our browser, we will have this result:

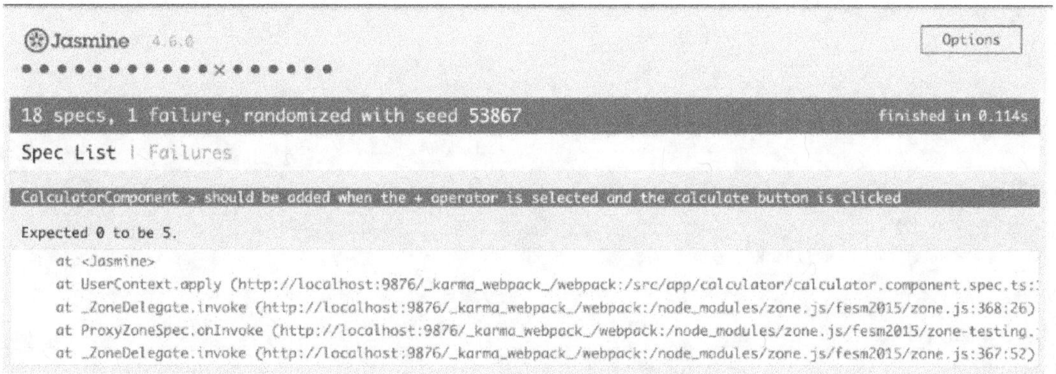

Figure 5.27 – Calculator operator (+) choice test failed in the browser

As we can see, our test has failed, and that's perfectly normal because the `calculate()` method is currently empty in our `CalculatorComponent`. We now need to add the minimum code required to make it functional. We need to take into consideration the previous functions contained in the class that we had to add in the previous chapters. Here's how it looks when we update the `calculate()` method in the `calculator.component.ts` file:

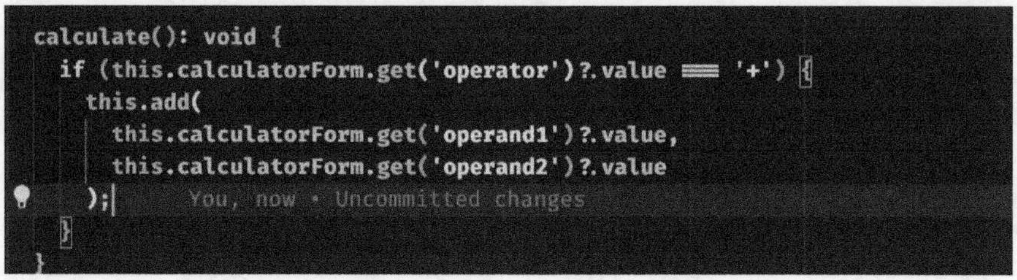

```
calculate(): void {
    if (this.calculatorForm.get('operator')?.value === '+') {
        this.add(
            this.calculatorForm.get('operand1')?.value,
            this.calculatorForm.get('operand2')?.value
        );
    }
}
```

Figure 5.28 – Calculator operator (+) choice

In our terminal, after running the ng test command, we will have this:

```
✓ Browser application bundle generation complete.
Chrome 119.0.0.0 (Mac OS 10.15.7): Executed 18 of 18 SUCCESS (0.114 secs / 0.091 secs)
TOTAL: 18 SUCCESS
```

Figure 5.29 – Calculator operator (+) choice test succeeded in the terminal

In our browser, we will have this result:

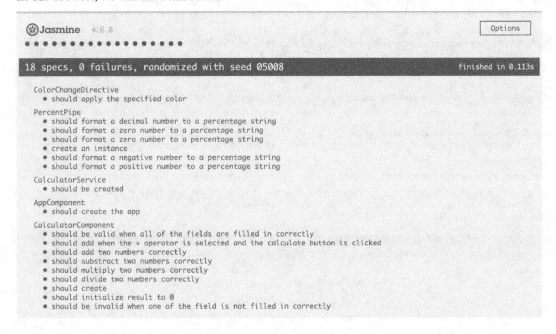

Figure 5.30 – Calculator operator choice (+) test succeeded in the browser

Now we'll do the same for the subtraction functionality. In our calculator.component.spec.ts file, we will add the following test suite:

```
it('should subtract when the - operator is selected and the calculate
button is clicked', () => {
    calculator.calculatorForm.get('operand1')?.setValue(2);
    calculator.calculatorForm.get('operand2')?.setValue(3);
    calculator.calculatorForm.get('operator')?.setValue('-');
  calculator.calculate();
  expect(calculator.result).toBe(-1);
});
```

In our terminal, after running the ng test command, we will have this:

```
✓ Browser application bundle generation complete.
Chrome 119.0.0.0 (Mac OS 10.15.7) CalculatorComponent should subtract when the - operator is selected and the calculate
button is clicked FAILED
        Expected 0 to be -1.
            at <Jasmine>
            at UserContext.apply (src/app/calculator/calculator.component.spec.ts:110:31)
            at _ZoneDelegate.invoke (node_modules/zone.js/fesm2015/zone.js:368:26)
            at ProxyZoneSpec.onInvoke (node_modules/zone.js/fesm2015/zone-testing.js:273:39)
            at _ZoneDelegate.invoke (node_modules/zone.js/fesm2015/zone.js:367:52)
Chrome 119.0.0.0 (Mac OS 10.15.7): Executed 19 of 19 (1 FAILED) (0.127 secs / 0.106 secs)
TOTAL: 1 FAILED, 18 SUCCESS
```

Figure 5.31 – Calculator operator choice (-) test failed in the terminal

In our browser, we will have this result:

```
⊛ Jasmine   4.6.0                                                                              Options

● ● ● ● ● ● ✕ ● ● ● ● ● ● ● ● ● ● ● ●

19 specs, 1 failure, randomized with seed 52354                                      finished in 0.125s

Spec List | Failures

CalculatorComponent > should subtract when the - operator is selected and the calculate button is clicked

Expected 0 to be -1.
    at <Jasmine>
    at UserContext.apply (http://localhost:9876/_karma_webpack_/webpack:/src/app/calculator/calculator.component.spec.ts:
    at _ZoneDelegate.invoke (http://localhost:9876/_karma_webpack_/webpack:/node_modules/zone.js/fesm2015/zone.js:368:26)
    at ProxyZoneSpec.onInvoke (http://localhost:9876/_karma_webpack_/webpack:/node_modules/zone.js/fesm2015/zone-testing.
    at _ZoneDelegate.invoke (http://localhost:9876/_karma_webpack_/webpack:/node_modules/zone.js/fesm2015/zone.js:367:52)
```

Figure 5.32 – Calculator operator choice (-) test failed in the terminal

Our test failed, and that's perfectly normal because our calculate() method doesn't currently handle subtraction in our CalculatorComponent. We now need to add the minimum code required to make it functional. We need to take into account the previous functions contained in the class that we had to add in the previous chapters. Here's how it looks when we update the calculate() method in the calculator.component.ts file:

```
if (this.calculatorForm.get('operator')?.value === '-') {
    this.substract(        You, now • Uncommitted changes
        this.calculatorForm.get('operand1')?.value,
        this.calculatorForm.get('operand2')?.value
    );
}
```

Figure 5.33 – Calculator operator choice (-)

In our terminal, after running the `ng test` command, we will have this:

```
✓ Browser application bundle generation complete.
Chrome 119.0.0.0 (Mac OS 10.15.7): Executed 19 of 19 SUCCESS (0.139 secs / 0.109 secs)
TOTAL: 19 SUCCESS
```

Figure 5.34 – Calculator operator choice (-) test succeeded in the terminal

In our browser, we will have this result:

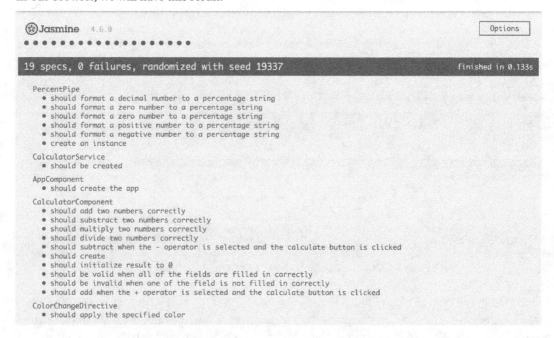

Figure 5.35 – Calculator operator choice (-) test succeeded in the browser

Next is the multiplication functionality. In our `calculator.component.spec.ts` file, we'll add this test suite:

```
it('should multiply when the * operator is selected and the calculate
button is clicked', () => {
    calculator.calculatorForm.get('operand1')?.setValue(2);
    calculator.calculatorForm.get('operand2')?.setValue(3);
    calculator.calculatorForm.get('operator')?.setValue('*');
calculator.calculate();
expect(calculator.result).toBe(6);
});
```

In our terminal, after running the ng test command, we will have this:

```
✓ Browser application bundle generation complete.
Chrome 119.0.0.0 (Mac OS 10.15.7) CalculatorComponent should multiply when the * operator is selected and the calculate
button is clicked FAILED
        Expected 0 to be 6.
            at <Jasmine>
            at UserContext.apply (src/app/calculator/calculator.component.spec.ts:118:31)
            at _ZoneDelegate.invoke (node_modules/zone.js/fesm2015/zone.js:368:26)
            at ProxyZoneSpec.onInvoke (node_modules/zone.js/fesm2015/zone-testing.js:273:39)
            at _ZoneDelegate.invoke (node_modules/zone.js/fesm2015/zone.js:367:52)
Chrome 119.0.0.0 (Mac OS 10.15.7): Executed 20 of 20 (1 FAILED) (0.149 secs / 0.134 secs)
TOTAL: 1 FAILED, 19 SUCCESS
```

Figure 5.36 – Calculator operator choice (*) test failed in the terminal

In our browser, we will have this result:

```
⊛ Jasmine    4.6.0                                                                    Options

● ● ● ● ● ● ● ● ● ● ● ● ✗ ● ● ● ● ● ● ● ● ●

20 specs, 1 failure, randomized with seed 38361                            finished in 0.147s

Spec List | Failures

CalculatorComponent > should multiply when the * operator is selected and the calculate button is clicked

Expected 0 to be 6.
    at <Jasmine>
    at UserContext.apply (http://localhost:9876/_karma_webpack_/webpack:/src/app/calculator/calculator.component.spec.ts:
    at _ZoneDelegate.invoke (http://localhost:9876/_karma_webpack_/webpack:/node_modules/zone.js/fesm2015/zone.js:368:26)
    at ProxyZoneSpec.onInvoke (http://localhost:9876/_karma_webpack_/webpack:/node_modules/zone.js/fesm2015/zone-testing.
    at _ZoneDelegate.invoke (http://localhost:9876/_karma_webpack_/webpack:/node_modules/zone.js/fesm2015/zone.js:367:52)
```

Figure 5.37 – Calculator operator choice (*) test failed in the browser

Our test failed, and that's perfectly normal because our calculate() method doesn't currently handle multiplication in our CalculatorComponent. We now need to add the minimum code required to make it functional. We need to take into account the previous functions contained in the class that we had to add in the previous chapters. Here's how it looks when we update the calculate() method in the calculator.component.ts file:

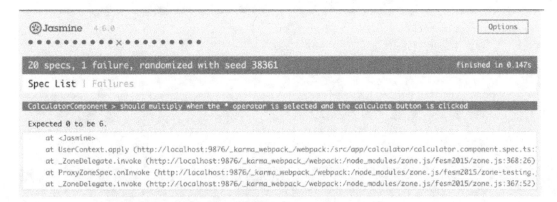

```
if (this.calculatorForm.get('operator')?.value === '*') {
  this.multiply(
    this.calculatorForm.get('operand1')?.value,
    this.calculatorForm.get('operand2')?.value
  );
}
```

Figure 5.38 – Calculator operator choice (*)

In our terminal, after running the ng test command, we will have this:

```
✓ Browser application bundle generation complete.
Chrome 119.0.0.0 (Mac OS 10.15.7): Executed 20 of 20 SUCCESS (0.112 secs / 0.092 secs)
TOTAL: 20 SUCCESS
```

Figure 5.39 – Calculator operator choice (*) test succeeded in the terminal

In our browser, we will have this result:

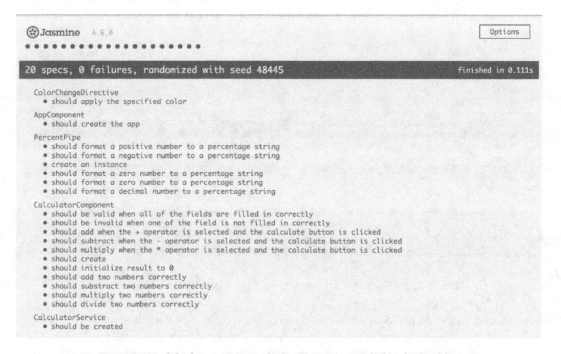

Figure 5.40 – Calculator operator choice (*) test succeeded in the browser

Finally, we'll set up the last operation, that is, divide. In our calculator.component.spec.ts file, we'll add this test suite:

```
it('should divide when the / operator is selected and the calculation
button is clicked.', () => {
    calculator.calculatorForm.get('operand1')?.setValue(3);
    calculator.calculatorForm.get('operand2')?.setValue(2);
    calculator.calculatorForm.get('operator')?.setValue('/');
    calculator.calculate();
    expect(calculator.result).toBe(1.5);
});
```

In our terminal, after running the ng test command, we will have this:

```
✓ Browser application bundle generation complete.
Chrome 119.0.0.0 (Mac OS 10.15.7) CalculatorComponent should divide when the / operator is selected and the calculation
button is clicked. FAILED
        Expected 0 to be 1.5.
            at <Jasmine>
            at UserContext.apply (src/app/calculator/calculator.component.spec.ts:126:31)
            at _ZoneDelegate.invoke (node_modules/zone.js/fesm2015/zone.js:368:26)
            at ProxyZoneSpec.onInvoke (node_modules/zone.js/fesm2015/zone-testing.js:273:39)
            at _ZoneDelegate.invoke (node_modules/zone.js/fesm2015/zone.js:367:52)
Chrome 119.0.0.0 (Mac OS 10.15.7): Executed 21 of 21 (1 FAILED) (0.115 secs / 0.101 secs)
TOTAL: 1 FAILED, 20 SUCCESS
```

Figure 5.41 – Calculator operator choice (/) test failed in the terminal

In our browser, we will have this result:

```
⊛ Jasmine   4.6.0                                                                              Options

• • • • • • • • • • • • ⨯ • • • • • • • •

21 specs, 1 failure, randomized with seed 69604                                    finished in 0.114s

Spec List  |  Failures

CalculatorComponent > should divide when the / operator is selected and the calculation button is clicked.

Expected 0 to be 1.5.
    at <Jasmine>
    at UserContext.apply (http://localhost:9876/_karma_webpack_/webpack:/src/app/calculator/calculator.component.spec.ts::
    at _ZoneDelegate.invoke (http://localhost:9876/_karma_webpack_/webpack:/node_modules/zone.js/fesm2015/zone.js:368:26)
    at ProxyZoneSpec.onInvoke (http://localhost:9876/_karma_webpack_/webpack:/node_modules/zone.js/fesm2015/zone-testing.
    at _ZoneDelegate.invoke (http://localhost:9876/_karma_webpack_/webpack:/node_modules/zone.js/fesm2015/zone.js:367:52)
```

Figure 5.42 – Calculator operator choice (/) test failed in the browser

Our test failed, and that's perfectly normal because our calculate() method doesn't currently handle division in our CalculatorComponent. We now need to add the minimum code required to make it functional. We need to take into account the previous functions contained in the class that we had to add in the previous chapters. Here's how it looks when we update the calculate() method in the calculator.component.ts file:

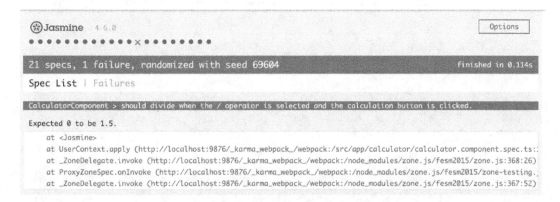

```
if (this.calculatorForm.get('operator')?.value === '/') {
  this.divide(
    this.calculatorForm.get('operand1')?.value,
    this.calculatorForm.get('operand2')?.value
  );
}
```

Figure 5.43 – Calculator operator choice (/)

In our terminal, after running the `ng test` command, we will have this:

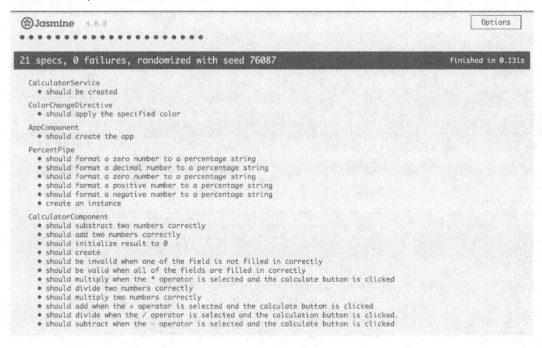

Figure 5.44 – Calculator operator choice (/) test succeeded in the terminal

In our browser, we will have this result:

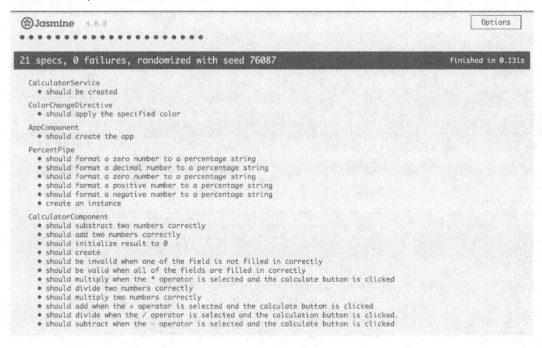

Figure 5.45 – Calculator operator choice (/) test succeeded in the browser

We can now run the `ng serve -o` command to see how our application behaves:

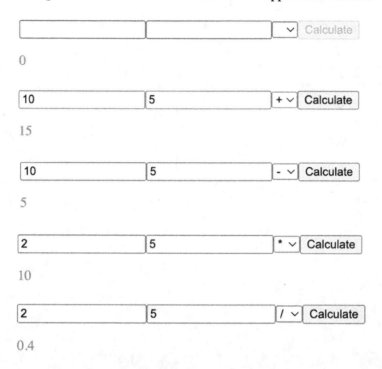

Figure 5.46 – Calculator user interface test

Summary

In this chapter, we learned how to test Angular pipes and apply TDD to reactive forms for reactive programming.

We explored the testing of Angular pipes, which are essential components used to transform input data for display in the view. The process involved creating test cases that covered different scenarios and edge cases to ensure the pipe behaved as expected.

We delved into the application of TDD to reactive forms in Angular. TDD is a software development approach that involves writing tests before the code, and these tests drive the development of the code. This approach ensures that the code is robust, reliable, and well tested.

In the next chapter, we will explore end-to-end testing with Protractor, Cypress, and Playwright.

Part 3:
End-to-End Testing

In this part, you'll get an overview of end-to-end testing with Protractor, Cypress, and Playwright. Then, you'll explore in depth how to use Cypress to write end-to-end tests.

This part has the following chapters:

- *Chapter 6, Exploring End-to-End Testing with Protractor, Cypress, and Playwright*

- *Chapter 7, Understanding Cypress and its Role in End-to-End Tests for Web Applications*

- *Chapter 8, Writing Effective End-to-End Component Tests with Cypress*

6

Exploring End-to-End Testing with Protractor, Cypress, and Playwright

In software development, **end-to-end** (**E2E**) testing plays a crucial role in ensuring the integration and functionality of an application from start to finish. E2E testing encompasses the entire application flow, simulating interactions with the real user to verify that the application functions as intended in various usage scenarios.

E2E testing offers a multitude of benefits that contribute to the overall quality and reliability of software applications. E2E testing identifies and eliminates usability issues that can hinder user satisfaction and adoption.

By thoroughly testing the application's integrated components, E2E testing reduces the likelihood of unexpected errors or disruptions during actual use. E2E testing enables early identification of defects, minimizing the time and costs associated with problem resolution later in the development cycle. Proactive fault detection with E2E testing can significantly reduce long-term maintenance costs.

Several powerful tools are available to facilitate E2E testing, each with its own strengths and features. We'll take a look at three major options: Protractor, Cypress, and Playwright.

In this chapter, we will explore the theory behind E2E testing, the benefits of E2E testing after long and deep analysis, and the philosophy behind some E2E tools such as Protractor, Cypress, and Playwright.

In summary, here are the main topics that will be covered in this chapter:

- Discovering E2E testing
- Analyzing the benefits of E2E testing in a project
- Exploring the different tools that can be used for E2E testing such as Protractor, Cypress, and Playwright

Technical requirements

To follow along with the examples and exercises in this chapter, you will need to have a basic understanding of Angular and TypeScript, as well as the following technical requirements:

- Node.js and npm installed on your computer
- Angular CLI installed globally
- A code editor, such as Visual Studio Code, installed on your computer

Understanding E2E testing

E2E testing is an in-depth method of software testing that evaluates the entire application flow from the user's point of view, ensuring that all components work as expected and that the software functions correctly in real-life scenarios. It encompasses testing of the application's user interface, backend systems, external APIs, and data interactions to ensure a seamless and consistent user experience. Unlike unit testing, which focuses on individual components, E2E testing examines the whole application as a unified entity, ensuring that all components work in harmony to deliver the desired functionality.

Discovering the benefits of E2E testing

E2E testing offers a myriad of benefits, making it an invaluable tool for software development teams:

- **Early detection of defects**: E2E testing enables defects to be detected early in the development cycle, reducing the cost and effort required to correct them later.

- **Enhanced user experience**: E2E testing is a method of ensuring that a software application works as intended from the user's point of view. This approach improves the overall user experience and minimizes the risk of dissatisfaction.

- **Increased confidence in software quality**: E2E testing inspires confidence in software quality, reducing the risk of unforeseen problems in production and preserving the reputation of the software development team.

- **Streamlined development process**: E2E testing can be automated, streamlining the development process, facilitating **continuous integration and continuous delivery (CI/CD)** practices, and enabling software to be released faster and more efficiently.

In the next section, we will learn how to implement E2E tests.

Exploring different approaches to E2E testing

E2E testing encompasses two main strategies: script-based testing and exploratory testing. Script-based testing relies on automated scripts to execute test cases, ensuring the consistency, repeatability, and efficiency of test procedures. On the other hand, exploratory testing is a practical approach that involves manually testing the application while exploring different scenarios. Let's take an in-depth look at each of the two approaches:

- **Script-based testing**: Script-based testing is a software testing methodology in which automated scripts are used to execute test cases. This approach guarantees the consistency, repeatability, and efficiency of test procedures, making it a reliable tool for ensuring that software applications run smoothly.

  There are several popular frameworks for script-based testing, including Protractor, Cypress, and Playwright. Each of these frameworks has its own features and benefits:

  - **Protractor** is an E2E testing framework for Angular and AngularJS applications. It aims to simplify the process of setting up tests, making them more readable and producing results that are easier to understand.

  - **Cypress** is a next-generation frontend testing tool designed for the modern web. It provides a complete testing solution, including the ability to write tests directly in JavaScript, without the need for additional preprocessors or compilers. Cypress is renowned for its ease of use and rapid installation.

  - **Playwright**, developed and maintained by Microsoft, is an open source, Node.js-based automation framework for E2E testing. It was created to meet the need for automated E2E testing across multiple browsers. Playwright's main objective is to run on the main browser engines – Chromium, WebKit, and Firefox. It offers extensive, native mobile testing capabilities, supporting mobile automation testing on Android and iOS platforms.

- **Exploratory testing**: Exploratory testing is a dynamic and flexible approach to software testing that involves the tester actively exploring the application and trying out different scenarios, inputs, and interactions to identify bugs and problems. Unlike script-based testing, which follows predefined test cases and steps, exploratory testing is based on the tester's curiosity, creativity, and intuition.

  This approach is particularly useful when there are many unknowns about exactly what the software is supposed to do or how it should behave in real-life situations.

  Exploratory testing is often used when the application is complex or constantly evolving. It is also useful when rapid feedback is needed, when requirements are unclear, or when time is of the essence.

There are different types of exploratory testing – free-style testing, scenario-based testing, and strategy-based testing:

- **Free-style testing**: This is useful when you need to quickly familiarize yourself with the application

- **Scenario-based testing**: This focuses on real-world usage scenarios

- **Strategy-based testing**: This combines exploratory testing with well-known testing methods

Exploratory testing can sometimes be more useful in identifying more subtle defects that formal testing might miss. However, it requires a high degree of skill and understanding of the application.

Comparison between script-based testing and exploratory testing

The comparison between script-based and exploratory testing provides us with several important insights:

- **Consistency and efficiency**: Script-based testing guarantees the consistency and repeatability of test procedures, which is crucial in large-scale projects where many tests are required. It can also be more efficient, as it eliminates the need for manual test execution.

- **Flexibility and discovery**: On the other hand, exploratory testing offers greater flexibility as it allows testers to freely explore the application. This can lead to the discovery of potential problems that scripted tests might miss, particularly in the case of complex applications or when the application undergoes frequent changes.

- **Documentation**: A significant difference between the two approaches is the level of documentation involved. Script-based testing typically involves detailed documentation of test cases and steps, while exploratory testing can involve less documentation, which can potentially lead to critical bugs being missed.

Limitations of script-based testing and exploratory testing

Both script-based testing and exploratory testing approaches have their limitations. For example, exploratory testing can be influenced by the tester's biases, and may fail to capture all potential problems if the tester lacks sufficient knowledge or fails to explore thoroughly. Similarly, scripted tests may miss obvious defects if the test cases do not cover all possible scenarios.

Many organizations prefer a mixed approach, combining elements of script-based and exploratory testing, to leverage the strengths of each method and mitigate their weaknesses.

Understanding the context of the project, for example, whether it's a small project or part of a larger one, can guide the choice between script-based and exploratory testing.

We've looked at some of the tools previously mentioned for E2E testing. We will now explore them in more detail.

Harnessing the power of E2E testing tools

A variety of tools are available to facilitate E2E testing, empowering software development teams to effectively implement this testing strategy.

Let's take a closer look at some of these tools.

Selenium – proven power for web application testing

Selenium is a renowned open source framework for automating web application testing. Its comprehensive feature set meets a wide range of testing needs, including functional, integration, and cross-browser testing. Selenium supports a variety of programming languages, including Java, Python, and C#, enabling testers to choose the language that matches their expertise and preferences. In addition, Selenium boasts extensive compatibility with the most popular browsers, such as Chrome, Firefox, Safari, and Edge. This broad compatibility ensures that tests can be run on different browsers, eliminating browser-specific discrepancies and guaranteeing a seamless user experience.

Cypress – simplicity and speed for web testing

Cypress has established itself as a JavaScript-based testing framework for web applications, renowned for its simplicity, speed, and ease of use. Unlike Selenium, which requires the installation of binary drivers and configuration, Cypress offers a standalone package that simplifies the installation process. Cypress's Node.js-based architecture delivers exceptional performance, enabling tests to run quickly, reducing test times, and improving overall efficiency. Its intuitive API and simple syntax make it accessible to testers of all levels, facilitating rapid test development and maintenance.

Appium – mobile application testing on multiple platforms

Appium is a versatile framework for automating mobile app testing on different platforms, including iOS and Android. It uses the Appium driver to communicate with native mobile apps, enabling testers to simulate user interactions such as tapping, swiping, and typing. Appium's cross-platform compatibility means that tests can be run on real devices or simulators, ensuring that the application works consistently in different environments.

Protractor – seamless automation for Angular applications

Protractor, designed specifically for Angular applications, leverages Selenium to provide a complete testing framework. Its integration with Angular framework components simplifies test development and maintenance, allowing testers to interact with elements and use Angular-specific functions. Protractor's support for local and remote test environments meets a wide range of testing needs.

Playwright – cross-browser testing with performance

Playwright, developed by Microsoft, has established itself as a powerful tool for cross-browser testing. It supports Chromium, WebKit, and Firefox browsers, enabling comprehensive tests to be carried out on different rendering engines. Playwright's Node.js-based architecture delivers exceptional performance, enabling tests to run quickly and efficiently. Its API is simple and intuitive, making it accessible to testers of all levels.

The choice of E2E testing tools depends on the specific needs and preferences of the software development team.

Here are the strengths of the E2E testing tools:

- Selenium's flexibility and broad browser compatibility make it a solid choice for comprehensive web application testing

- Cypress's simplicity and speed appeal to teams looking for a streamlined approach to web testing

- Appium meets mobile testing needs, enabling automation on iOS and Android platforms

- Protractor facilitates seamless automation of Angular applications

- Playwright offers high-performance, cross-browser testing

Ultimately, the most effective tool is the one that matches the team's expertise, project requirements, and testing objectives.

In the next section, we will analyze the benefits of E2E testing in a project.

Analyzing the benefits of E2E testing in a project

The adoption of E2E testing as part of software development projects reveals a multitude of benefits that improve the quality and user experience of software products.

While E2E tests offer a multitude of advantages, their implementation is not without its problems. Understanding these obstacles is essential to developing effective strategies for overcoming them:

- **Complexity**: E2E tests often have a labyrinthine structure, which makes them difficult to create and maintain. This complexity stems from the need to test multiple application components and accurately simulate the actions of a real user.

- **Scaling**: E2E tests are prone to instability, a characteristic that manifests itself in intermittent failures for reasons that can be difficult to decipher. This intermittence can compromise the reliability of E2E test results.

- **Slowness**: E2E tests often run at a slow pace due to their global nature, which involves interactions with the whole application and potential delays in waiting for responses from external systems.

Despite the inherent challenges, E2E testing remains an invaluable tool for improving software quality. By adopting proven strategies, developers and testers can effectively meet these challenges and reap the many benefits of E2E testing:

- **Get involved early**: It's vital to launch E2E testing early in the development cycle. This proactive approach facilitates the early detection of defects, making their treatment considerably easier and more cost-effective.

- **Leverage automation**: Automation strengthens E2E testing by streamlining the execution process and minimizing the expenditure of time and effort. Automation also facilitates the seamless integration of E2E tests into the CI/CD pipeline.

- **Meticulous test design**: Meticulous care in the design of E2E tests is essential to ensure their effectiveness and efficiency. This means focusing on testing the most critical user flows and identifying the most likely points of failure.

- **Leverage specialized tools**: There is a plethora of tools available to support E2E testing. It is essential to select a tool that matches the project's requirements and offers user-friendly functionality.

In the next section, we'll explore the different tools that can be used for E2E testing, such as Protractor, Cypress, and Playwright.

Exploring Protractor, Cypress, and Playwright for E2E testing

There are many different E2E testing tools available and choosing the right one can be tricky. This review examines three popular tools: Protractor, Cypress, and Playwright. We'll discover their strengths, weaknesses, and ideal use cases, enabling you to select the ideal tool for your E2E tests.

Protractor

Produced by Google's team of developers, Protractor was first designed for Angular applications, then made available as an open source solution. Today, it extends its capabilities beyond Angular, adapting to non-Angular applications as well. It is an enhanced version of WebDriver.js, incorporating all Selenium WebDriver features plus specialized functions for Angular development.

What is Protractor?

Protractor is an open source testing framework primarily used for E2E testing of Angular and AngularJS applications. Although originally designed for Angular, it has evolved to support testing of Angular and non-Angular web applications. It works in the same way as a real user, running tests in a real web browser and resulting in realistic test scenarios. Protractor is a Node.js application and acts as a wrapper around Selenium WebDriver using WebDriverJS, which is the JavaScript binding for the Selenium WebDriver API. It can create tests that interact with a real browser, enabling E2E testing.

It integrates with other technologies such as Node.js, Jasmine, Selenium, Mocha, and others. It is particularly well-suited to Angular applications due to its origins in the Angular team and its use of Angular-specific locators to identify DOM elements. However, as of November 2022, Protractor is considered obsolete, and the end of development was scheduled for the end of 2022. The team behind Protractor announced this decision due to changes in web development and the JavaScript language, which make Protractor less compatible with contemporary applications.

Why choose Protractor?

Protractor provides a set of Angular-specific APIs that make it easier to interact with Angular elements and perform actions such as waiting for Angular processes to complete, handling asynchronous operations, and managing synchronization with the Angular event loop.

You can simulate real user interactions such as clicking buttons, filling out forms, navigating between pages, and verifying the behavior of Angular elements. This makes Protractor a powerful tool for E2E testing of Angular applications.

It automatically waits for Angular processes to complete before executing the next step, eliminating the need for explicit waits and reducing flakiness in tests. This makes it a reliable choice for testing Angular applications.

It supports various locators specifically designed for Angular applications, including model, binding, repeater, and CSS selectors. These locators make it easier to identify and interact with Angular elements, which is particularly useful for testing Angular applications.

It allows you to run tests in multiple browsers, including Chrome, Firefox, and Safari, enabling cross-browser testing of your Angular applications. This feature ensures that your Angular application works correctly across different browsers, which is crucial to a good user experience.

Integrating it with a CI system allows you to automate the execution of tests on every code commit, ensuring that your Angular application remains stable and functional. This feature is particularly beneficial to large teams and projects where multiple developers are working on the same code base.

Features of Protractor testing

By providing built-in support for asynchronous operations via callbacks, promises, and async/await, Protractor empowers developers to do the following:

- **Boost test execution speed**: Achieve faster E2E testing, especially for Angular web apps with dynamic elements
- **Enhance test readability**: Write cleaner and more maintainable tests with modern asynchronous programming methods
- **Increase test reliability**: Simplify handling dynamic behaviors and improve test stability

One of the key features of Protractor is its Automatic Waiting feature. Here's how it works:

- **Implicit waiting**: By default, Protractor assumes a short wait time (usually around 10 seconds) before throwing an error if an element isn't found. This eliminates the need for manual `wait` statements between every action in your tests.

- **Angular synchronization**: Protractor integrates with Angular's framework to understand when the application has finished processing and is ready for interaction. This ensures your tests don't try to interact with elements before they're fully loaded and available.

While Protractor handles most waiting scenarios automatically, there are situations where you might still need explicit waits:

- **Custom conditions**: You can use Protractor's `ExpectedConditions` API to define specific conditions for waiting, such as waiting for an element to be clickable or for a specific text to appear.

- **Long waits**: The default implicit wait time might be too short for certain scenarios. You can configure a custom wait time for specific test cases.

    Protractor supports both Angular and non-Angular applications. It provides combined E2E testing for web applications built using AngularJS. This makes it a versatile tool for testing a wide range of web applications.

Protractor supports the Page Objects pattern, which is a design pattern that enhances test maintenance and reduces code duplication. With this pattern, you can create reusable page objects that can be used across different tests.

Cypress

Designed for web development and test automation, Cypress is a JavaScript-based E2E testing framework that streamlines both web and API testing.

What is Cypress?

Cypress is an open source, JavaScript-based E2E testing framework specifically designed for modern web applications. It offers a developer-friendly approach to automating web UI tests, streamlining the process for both frontend developers and QA engineers.

Why choose Cypress?

Here are some key advantages of using Cypress for your web application testing:

- **Simplified setup**: Compared to other tools such as Selenium, Cypress requires minimal configuration. It often involves just installing the package and writing tests in JavaScript.

- **Fast and efficient**: Cypress runs tests directly in the browser, eliminating the need for separate WebDriver setup and browser launching. This makes tests execute significantly faster, improving development and testing workflows.

- **Automatic waiting**: Cypress handles waiting for elements to load and become interactive automatically. You don't need to write explicit waits or sleeps into your tests, reducing code complexity and making tests more robust.

- **Easy debugging**: Cypress provides a visual debugger that allows you to step through your tests, inspect element properties, and identify issues quickly. This streamlines the debugging process.

- **Time-travel debugging**: With Cypress's time-travel debugging feature, you can rewind or fast-forward a test execution to pinpoint the exact moment when a failure occurs, making troubleshooting more efficient.

- **Cross-browser compatibility**: Cypress supports testing across different browsers (Chrome, Firefox, Edge, etc.) out of the box. You can configure tests to run on various browsers or use a CI platform to automate browser testing.

- **Integration with development tools**: Cypress integrates seamlessly with popular developer tools such as DevTools, allowing you to leverage existing debugging skills within the testing environment.

- **Active community and support**: Cypress has a large and active community offering extensive documentation, tutorials, and support resources. This makes it easier to learn it, use it, and get help when needed.

Features of Cypress testing

Cypress boasts a rich set of features that contribute to its effectiveness in web application testing:

- **Command chain API**: Provides a natural way to write tests using JavaScript syntax, making them readable and maintainable

- **Automatic assertions**: Simplifies checking for expected element properties and behavior, ensuring your tests verify the correct functionality

- **Screenshots and video recordings**: Enables capturing screenshots or recording videos during tests, which can be helpful for visualizing errors and debugging issues

- **Network mocking**: Allows simulating server responses, facilitating testing of API interactions, edge cases, and various network scenarios

- **Custom commands**: Grants the ability to create reusable code blocks for frequently used testing actions, promoting code reusability and modularity
- **Integration testing**: Cypress can be used for both unit and integration testing of web applications, offering comprehensive testing coverage

By leveraging Cypress's strengths, you can build a robust and efficient test suite that helps ensure the quality and reliability of your web applications.

Playwright

Playwright is a modern, versatile, and powerful browser automation framework created by Microsoft. It enables developers to write automated tests, scrape web data, and interact with web applications using a single API that works seamlessly across multiple browsers, including Chrome, Firefox, and Microsoft Edge.

What is Playwright?

Playwright is a powerful, open source Node.js library for automating and testing web applications on various browsers such as Chrome, Firefox, Safari, and Edge. It can also automate browser actions such as clicks, form fill-ups, and page navigation. Playwright is an intuitive platform supporting several programming languages such as JavaScript, TypeScript, Python, and Java. The defining feature of parallelism is that you can test for multiple browsers simultaneously without the requirement of writing individual tests.

Why choose Playwright?

Playwright supports testing across multiple browsers, including Chromium, Firefox, and WebKit. This ensures cross-browser compatibility and a consistent user experience.

Playwright is known for its speed and efficiency. Its modern architecture and efficient browser automation result in faster test execution compared to other frameworks, reducing the overall time spent on testing.

Playwright's auto-wait API and advanced selector engine significantly improve test stability, reducing the likelihood of flaky tests and minimizing the need for manual intervention.

Playwright is built to handle modern web technologies, such as **single-page applications** (**SPAs**), better than other frameworks. It provides native support for interacting with dynamic content, making writing tests for complex applications easier.

Unlike some other frameworks that only support specific browsers, Playwright offers a consistent API for testing across multiple browsers, ensuring broader compatibility and a more comprehensive testing suite.

Playwright has built-in mocking that allows you to write minified E2E tests at the component level. This can be particularly useful for testing complex components without having to interact with the entire application.

Playwright supports powerful pseudo-CSS selectors that replace the only use cases for XPath. This can make your tests more readable and maintainable.

Playwright is about 30% faster than Selenium running the same tests due to its ability to create test contexts in 100 ms.

Playwright requires fewer dependencies and has a more straightforward installation process compared to Selenium. This means you can start writing tests more quickly and with less hassle.

Features of Playwright testing

Let's look at the key features of Playwright testing:

- Playwright provides consistent and reliable automation across different browsers, ensuring your web applications work well on all major platforms.

- Playwright is known for its speed and efficiency, making it a great choice for testing and automating web interactions.

- You can run Playwright scripts in headless mode (without a visible browser UI) for faster execution, or in headful mode for debugging and interaction.

- Playwright offers support for popular programming languages such as TypeScript, Python, and Java. This means you can write automation scripts in the language you're most comfortable with.

- Playwright uses native browser automation APIs to simulate user interactions accurately, resulting in more reliable tests.

You can leverage the full power of browser DevTools with Playwright, making it easier to inspect, debug, and diagnose issues in your web applications.

Summary

In this chapter, we have learned about E2E testing, an essential software development practice that ensures an application works perfectly from start to finish. E2E testing involves all integrated systems, both internal and external, to identify dependencies and verify the smooth flow of information.

We also learned about the most common E2E testing tools, Protractor, Cypress, and Playwright. We explored the strengths of each tool:

- **Protractor**: Designed for Angular applications, offering Angular-specific APIs for interacting with elements and handling asynchronous operations

- **Cypress**: A user-friendly tool that supports asynchronous/await syntax and concurrent test execution (albeit with limitations in the free version)

- **Playwright**: Provides a modern async/await approach to test scripting, encourages simultaneous test execution for faster execution, and boasts a simple API for efficient test development

Understanding the advantages and limitations of these tools enables us, as a developer, to make informed decisions and select the tool that's best suited to the specific needs of your project, taking into account the technology stack and team familiarity.

The following chapter takes a closer look at Cypress and its role in the E2E testing of web applications. This focused summary reinforces the chapter's key learnings, highlights its value (learning about E2E testing and tools), and provides an overview of the topic of the next chapter.

7

Understanding Cypress and its Role in End-to-End Tests for Web Applications

In the world of web development, Angular has emerged as one of the most popular frameworks for building dynamic and robust web applications. With its extensive features and powerful capabilities, Angular enables developers to create seamless user experiences. However, ensuring the reliability and functionality of these applications requires thorough testing, and this is where **Cypress** comes into play.

Cypress is a versatile **end-to-end** (**E2E**) testing framework that seamlessly integrates with Angular projects, allowing developers to write and execute component tests that cover the entire application flow. In this chapter, we will explore the role of Cypress in E2E testing for Angular projects and guide you through the process of discovering, setting up, and writing your first component test with Cypress.

We will begin by discovering Cypress and understanding its unique features and advantages. From its intuitive API to its real-time reloading and debugging capabilities, Cypress provides developers with a seamless and efficient testing experience. By grasping the core concepts of Cypress and its integration with Angular, developers can harness its power to ensure the quality and reliability of their Angular applications.

Next, we will dive into the process of setting up Cypress in an Angular project. We will explore the necessary configurations and dependencies, guiding you through the steps to integrate Cypress seamlessly into your Angular development workflow. With a solid foundation in place, you will be ready to leverage the full potential of Cypress in your Angular projects.

Finally, we will guide you through the process of writing your first E2E component test with Cypress in an Angular project. We will cover the essential aspects of writing effective component tests, including selecting elements, interacting with the application, and asserting expected behavior.

In summary, here are the main topics that will be covered in this chapter:

- Discovering Cypress and its role in an Angular project

- Setting up Cypress in an Angular project

- Writing your first E2E component test with Cypress in an Angular project

Technical requirements

To follow along with the examples and exercises in this chapter, you will need to have a basic understanding of Angular and TypeScript, as well as the following technical requirements:

- Node.js and **Node Package Manager** (**npm**) installed on your computer

- Angular CLI installed globally

- A code editor, such as Visual Studio Code, installed on your computer.

The code files of this chapter can be found at `https://github.com/PacktPublishing/Mastering-Angular-Test-Driven-Development/tree/main/Chapter%207`.

Discovering Cypress and its role in an Angular project

Testing plays a vital role in ensuring the quality and reliability of applications. When it comes to Angular projects, developers need a robust testing framework that seamlessly integrates with the framework's components and provides an efficient and comprehensive testing experience. This is where Cypress, an E2E testing framework, comes into the picture. In this section, we will explore Cypress and its role in Angular projects, uncovering its unique features and advantages.

Understanding Cypress

Cypress is a powerful JavaScript-based testing framework that allows developers to write and execute E2E tests for web applications. What sets Cypress apart from other testing frameworks is its ability to run tests directly in the browser, enabling real-time reloading and debugging. This feature makes Cypress an ideal choice for Angular projects, where developers can test their applications in a live environment, mimicking real user interactions.

Seamless integration with Angular

Cypress seamlessly integrates with Angular projects, making it an excellent choice for testing Angular applications. It leverages Angular's powerful components, directives, and services, allowing developers to write tests that cover the entire application flow. Whether it's testing individual components, validating complex user flows, or ensuring cross-browser compatibility, Cypress provides the necessary tools and capabilities to accomplish these tasks effectively.

Efficient testing workflow

One of the key advantages of Cypress is its intuitive API, which simplifies the process of writing tests. With its declarative syntax, developers can easily select elements, interact with the application, and assert expected behavior. Cypress also offers a comprehensive set of built-in commands and assertions, making it easier to write expressive and readable tests. Furthermore, Cypress provides powerful debugging features, allowing developers to inspect elements, step through the code, and troubleshoot issues in real time.

Real-time reloading and debugging

Cypress's real-time reloading and debugging capabilities are particularly valuable in Angular projects. As developers make changes to their code, Cypress automatically reloads the application, reflecting the updates in real time. This feature significantly speeds up the testing process, as developers can instantly see the impact of their changes without the need to manually refresh the browser. Additionally, Cypress's powerful debugging tools enable developers to pinpoint and resolve issues quickly, improving the overall efficiency of the testing workflow.

In the next section, we'll learn how to configure Cypress in our Angular project.

Setting up Cypress in our Angular project

Now that we've understood what Cypress is for and its role in an Angular project, we'll look at how to configure it in our project.

Installing Cypress

To begin, you need to have Node.js and npm installed on your machine. If you don't have them already, download and install them from the official Node.js website.

Once you have Node.js and npm installed, open your terminal and navigate to your Angular project's root directory. Run the following command to install Cypress as a dev dependency:

```
$ npm install cypress --save-dev
```

The previous command will download and install Cypress in your project. This is how it works in the terminal:

```
→  getting-started-angular-tdd git:(main) x npm i cypress --save-dev

added 109 packages, and audited 1011 packages in 9s

115 packages are looking for funding
  run `npm fund` for details

5 vulnerabilities (4 moderate, 1 critical)

To address all issues, run:
  npm audit fix

Run `npm audit` for details.
```

Figure 7.1 – Cypress installation

After installation, you can see the dependency in the package.json file:

```
"devDependencies": {
  "@angular-devkit/build-angular": "^16.1.3",
  "@angular/cli": "~16.1.3",
  "@angular/compiler-cli": "^16.1.0",
  "@types/jasmine": "~4.3.0",
  "cypress": "^13.6.2",
  "jasmine-core": "~4.6.0",
  "karma": "~6.4.0",
  "karma-chrome-launcher": "~3.2.0",
  "karma-coverage": "~2.2.0",
  "karma-jasmine": "~5.1.0",
  "karma-jasmine-html-reporter": "~2.1.0",
  "typescript": "~5.1.3"
}
```

Figure 7.2 – Cypress dependency in package.json

In the next section, we'll look at how to configure Cypress.

Configuring Cypress

As previously mentioned, we'll take a step-by-step look at how to configure Cypress:

1. After installing Cypress, run the following command in your terminal from your project root:

    ```
    $ npx cypress open
    ```

```
→  getting-started-angular-tdd git:(main) ✗ npx cypress open

It looks like this is your first time using Cypress: 13.6.2

✓  Verified Cypress! /Users/intelligencia/Library/Caches/Cypress/13.6.2/Cypre…

Opening Cypress ...
```

Figure 7.3 – Open Cypress in the command line

You will now see your browser launch and it will present the interface shown in *Figure 7.4*:

Figure 7.4 – Cypress launch interface

2.   Click on **Continue**, and you'll be redirected to the following interface:

Figure 7.5 – Cypress E2E test interface of preference

3.   Choose **E2E Testing** and you'll now be redirected to the interface shown in *Figure 7.6*:

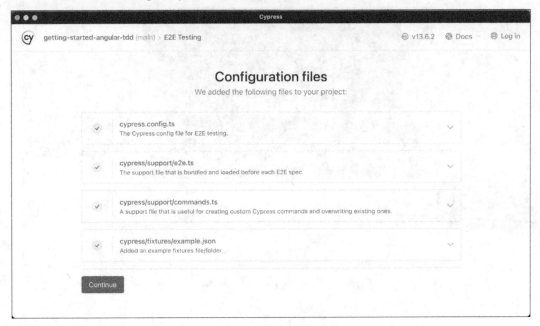

Figure 7.6 – Cypress E2E testing control panel

These files are part of a Cypress testing setup for a web application. Let's break down what each file does:

- `cypress.config.ts`: The `cypress.config.ts` file allows you to customize Cypress to fit the needs of your project, including setting up proxy configurations, configuring network requests, and specifying custom command line flags.

- `cypress/support/e2e.ts`: This file typically contains global commands and utilities that are available across all your E2E tests. It's a place where you can define functions that perform common actions, like logging in, navigating between pages, or interacting with elements in a consistent way. By defining these commands here, you ensure they're available in every test file without needing to redefine them.

- `cypress/support/commands.ts`: Similar to `e2e.ts`, the `commands.ts` file is used to extend Cypress's built-in commands with custom ones. This file is specifically focused on adding new commands that can be used within your test suites. These commands can encapsulate complex interactions or sequences of actions that you find yourself repeating across multiple tests. Defining custom commands helps keep your tests **DRY (Don't Repeat Yourself)**, making them easier to maintain and understand.

- `cypress/fixtures/example.json`: The fixtures folder is used to store JSON files that contain data used in your tests. These could be mock API responses, sample data for seeding databases, or any other static data required for your tests to run. The `example.json` file would be one such fixture file, containing example data that your tests might need to interact with.

4. After clicking on **Continue**, you will see this interface to choose your preferred browser for E2E testing, as shown in *Figure 7.7*:

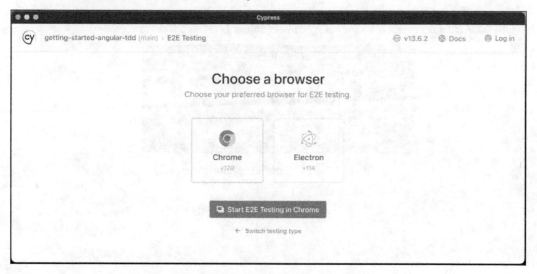

Figure 7.7 – The browser selection interface in Cypress

5. Once all the configuration has been completed, you should be able to access the interface, as shown in *Figure 7.8*:

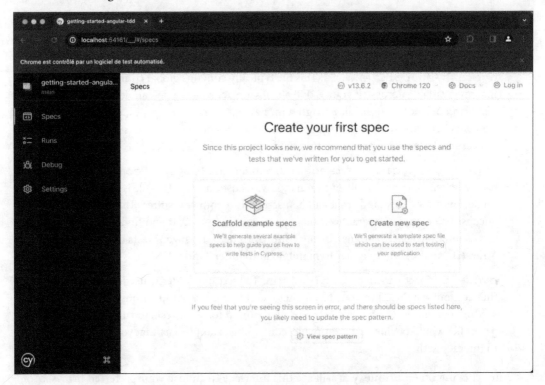

Figure 7.8 – Cypress E2E testing dashboard interface

6. In the Angular project directory, a `cypress` folder is created like this:

Figure 7.9 – The Cypress support folder

7.   Finally, we're going to add this script to `package.json` to avoid making `npx cypress open` the next few times:

```
"scripts": {
  "ng": "ng",
  "start": "ng serve",
  "build": "ng build",
  "watch": "ng build --watch --configuration development",
  "test": "ng test",
  "cypress:open": "cypress open"
```

Figure 7.10 – Cypress command in package.json scripts

In the next section, we'll take a look at how to write our first E2E test.

Writing your first E2E test

Once Cypress has been properly configured for E2E testing, we can get down to business. Let's take a look at how to write our first E2E test:

1.   Now that Cypress is installed and configured, you can start writing your first test. You can click on **Create new spec** in this interface as shown in *Figure 7.11*:

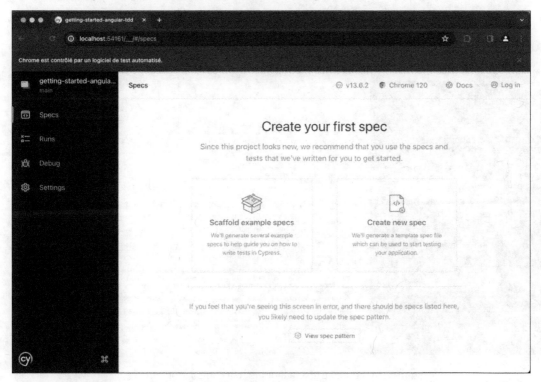

Figure 7.11 – Spec creation interface in Cypress

2.   After doing that, you can see in *Figure 7.12* a new file named `spec.cy.ts`, created in the e2e folder of your project's `cypress` folder:

Figure 7.12 – Spec file

This file contains the following:

```
1  describe('template spec', () => {
2    it('passes', () => {
3      cy.visit('https://example.cypress.io')
4    })
5  })
```

Figure 7.13 – The source code of our spec file

3.   Now, you may be faced with a surprise when you return to the browser where the E2E test is launched as shown in *Figure 7.14*:

Figure 7.14 – The E2E test failed due to a compilation error

4. To solve this issue, you can just remove `sourceMap: true` from your `tsconfig.json` file and you'll get this:

```json
{
  "compileOnSave": false,
  "compilerOptions": {
    "baseUrl": "./",
    "outDir": "./dist/out-tsc",
    "forceConsistentCasingInFileNames": true,
    "strict": true,
    "noImplicitOverride": true,
    "noPropertyAccessFromIndexSignature": true,
    "noImplicitReturns": true,
    "noFallthroughCasesInSwitch": true,
    "declaration": false,
    "downlevelIteration": true,
    "experimentalDecorators": true,
    "moduleResolution": "node",
    "importHelpers": true,
    "target": "ES2022",
    "module": "ES2022",
    "useDefineForClassFields": false,
    "lib": [
      "ES2022",
      "dom"
    ]
  },
  "angularCompilerOptions": {
    "enableI18nLegacyMessageIdFormat": false,
    "strictInjectionParameters": true,
    "strictInputAccessModifiers": true,
    "strictTemplates": true
  }
}
```

As you can see now, it works:

Figure 7.15 – The interface that displays all specs files

5.   Then, you can click on the spec.cy.ts file and if all goes well, you'll get this:

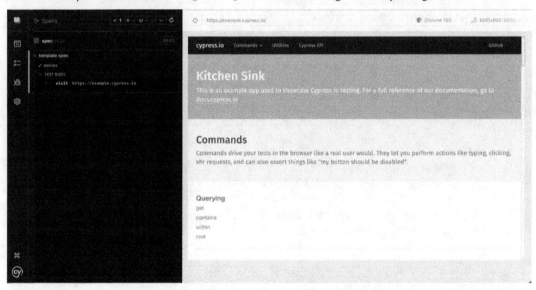

Figure 7.16 – The successful completion of our E2E test

Congrats! You've just written your first E2E test! Technically, we've generated it with the ease of Cypress.

Now that we know how to write an E2E test, the goal of the chapter has been achieved. We'll summarize everything we've covered in the next section.

Summary

This chapter covered discovering Cypress, configuring it, and taking it into an Angular project in the writing of an E2E test.

The chapter started with an introduction to the Cypress tool and explained why it's the most widely used tool in the Angular community, then showed how to configure Cypress in an Angular project, and finally taught you how to write your first E2E test with Cypress in Angular step by step.

In the next chapter, we will go into more depth with the writing of E2E tests with Cypress in a test-driven development approach, and we will improve and refactor our previous tests related to our components while respecting the good practices with Cypress.

8

Writing Effective End-to-End Component Tests with Cypress

In the ever-evolving world of web development, **end-to-end** (**E2E**) testing plays a crucial role in ensuring the reliability and robustness of web applications. E2E testing simulates real-world user scenarios, covering all aspects of an application, including user interfaces, APIs, databases, and other integrations.

One of the tools that developers often use for E2E testing is **Cypress**. Cypress is a freely available comprehensive testing solution for web applications, created with the aim of simplifying and optimizing the testing workflow for developers. What sets Cypress apart from other testing frameworks is its ability to execute tests directly in the browser, offering developers enhanced control and insight into the application being tested.

In this chapter, we will focus on how to write effective E2E tests for a calculator application using Cypress. We will explore the basics of Cypress, the structure of tests, and some advanced techniques. Our aim is to guide developers in writing comprehensive E2E tests that can effectively verify the functionality of a calculator app, ensuring that it behaves as expected under different conditions.

In summary, here are the main topics that will be covered in this chapter:

- Structuring E2E tests
- Writing E2E test cases
- Using Cypress custom commands

Technical requirements

To follow along with the examples and exercises in this chapter, you will need to have a basic understanding of Angular and TypeScript, as well as the following technical requirements:

- Node.js and npm installed on your computer
- The Angular CLI installed globally
- A code editor, such as Visual Studio Code, installed on your computer

The code files of this chapter can found at `https://github.com/PacktPublishing/Mastering-Angular-Test-Driven-Development/tree/main/Chapter%208`.

Structuring E2E tests

Cypress tests are structured using the `describe()`, `context()`, `it()`, and `specify()` functions, which are borrowed from Mocha (Mocha is a feature-rich JavaScript testing framework that runs on Node.js and in the browser, designed to make asynchronous testing simple and enjoyable. It is highly regarded for its versatility in testing applications across both the front- and backends, offering a wide range of benefits to developers). The `describe()` function is used to group related tests, `context()` is similar to `describe()`, and `it()` and `specify()` are used to write individual test cases.

We're going to structure our tests around the functionality of our calculator application. Our calculator performs four operations: *addition*, *subtraction*, *multiplication*, and *division*. In our `e2e` folder in the `cypress` folder, we'll create a file called `calculator.cy.ts`:

Figure 8.1 – The calculator.cy.ts file in e2e in the cypress folder

Once the file has been created, we will add these lines of code, which we will explain later:

```
describe('Calculator Functionality', () => {
context('Addition', () => {
it('adds two positive numbers correctly', () => {
// Test case for addition
});

it('adds two negative numbers correctly', () => {
// Test case for addition
});
});

it('add one positive number and one negative number correctly', () =>
{
// Test case for addition
});
});
```

```
context('Subtraction', () => {
it('subtracts two positive numbers correctly', () => {
// Test case for subtraction
});

it('subtracts two negative numbers correctly', () => {
// Test case for subtraction
});

it('subtracts one positive number and one negative number correctly',
() => {
// Test case for subtraction
});
});
});
```

The preceding lines of code begin with the definition of a test suite called `Calculator Functionality`. A **test suite** is a set of related tests. In this case, all the tests contained in the description block are linked to the functionality of a calculator.

The `context` function is used to group together related tests. Here, it groups all tests related to the calculator's `Addition` and `Subtraction` functions.

The `it` function defines an individual test scenario. Here, it defines a test case for the addition of two positive numbers or two negative numbers and their subtraction.

Now, we'll add the other operations (namely, `Multiplication` and `Division`), and we'll finalize the implementation of our test suite:

```
context('Multiplication', () => {
it('multiplies one positive number and zero correctly', () => {
// Test case for multiplication
});

it('multiplies two positive numbers correctly', () => {
// Test case for multiplication
});

it('multiplies two negative numbers correctly', () => {
// Test case for multiplication
});

it('multiplies one positive number and one negative number correctly',
() => {
```

```
// Test case for multiplication
});
});

context('Division', () => {
it('divides a positive non-zero number by another positive non-zero
number', () => {
// Test case for division
});

it('divides a negative non-zero number by another positive non-zero
number', () => {
// Test case for division
});

it('divides a negative non-zero number by another negative non-zero
number', () => {
// Test case for division
});

it('divides a positive non-zero number by another negative non-zero
number', () => {
// Test case for division
});

it('divides a positive non-zero number by zero', () => {
// Test case for division
});

it('divides a negative non-zero number by zero', () => {
// Test case for division
});

it('divide zero by zero', () => {
// Test case for division
});
});
```

This code sets up a structure for testing the addition, subtraction, multiplication, and division functionalities of a calculator application. Each operation has its own group of tests, and each test case describes a specific scenario.

In our browser, we have the following result:

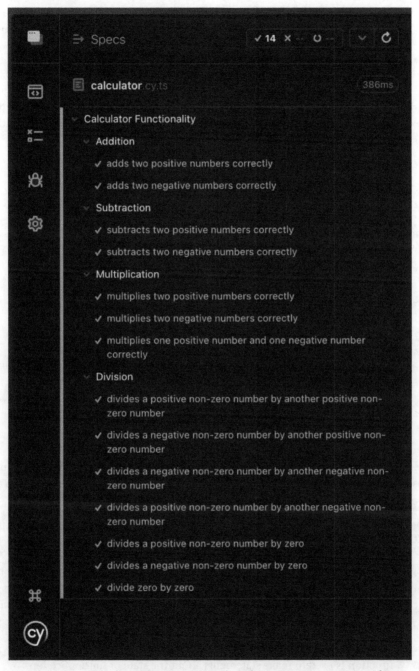

Figure 8.2 – Calculator E2E tests succeeded after setting up a structure for
testing addition, subtraction, multiplication, and division

The result in the browser proves that our test cases written in the E2E context are without errors. However, we haven't yet written the corresponding logic in our test cases.

In the next section, we'll now write the corresponding tests for each context.

Writing test cases

When writing test cases, you should try to cover different scenarios, boundary cases, and potential errors. For a calculator application, you may want to test addition, subtraction, multiplication, and division operations under normal conditions and in borderline cases (such as *division by zero*).

Let's complete our different contexts with our different tests.

Addition context

In this section, we'll look at the various E2E test cases related to addition.

Adds two positive numbers correctly

In this test case, we will see how to write the E2E test to sum two positive numbers:

```
it('adds two positive numbers correctly', () => {
cy.visit('http://localhost:4200/');
cy.get('input').first().type('5');
cy.get('select').select('+').should('have.value', '+');
cy.get('input').last().type('3');
cy.get('button').click();
cy.get('p').should('have.text', '8');
});
```

In the preceding code, the `it()` function is used to define an individual test case. The first argument is a string describing what the test case is supposed to do. In this case, it is `'adds two positive numbers correctly'`.

The `cy.visit()` command is used to visit a URL. Here, it's the URL that leads to our calculator's user interface.

The following line is used to enter `'5'` in the calculator's first input field. The `cy.get()` function is used to obtain elements from the DOM. The `first()` function is used to obtain the first element of the set of corresponding elements. The `type()` command is used to type into a text input field.

Then, select the + operator from the drop-down menu on our user interface and check that the selected value is indeed `'+'`. The `select()` command is used to select an option from a drop-down menu. The `should()` function is used to make statements about the state of the application.

The following line is used to enter '3' in the calculator's first input field. The cy.get() function is used to obtain DOM elements. The last() function is used to obtain the last element of the set of corresponding elements. The type() command is used to type into a text input field.

Then, click on the button to perform the calculation. The click() command is used to simulate a mouse click.

Finally, the last line of code checks that the result of the calculation ('8') is displayed correctly. The should() function is again used to make this assertion.

In our browser, we have this result:

Figure 8.3 – The "adds two positive numbers correctly" E2E test succeeded

In the next section, we'll learn how to correctly add two negative numbers.

Adds two negative numbers correctly

In this test case, we will see how to write the E2E test to add two negative numbers:

```
it('adds two negative numbers correctly', () => {
cy.visit('http://localhost:4200/');
cy.get('input').first().type('-5');
cy.get('select').select('+').should('have.value', '+');
cy.get('input').last().type('-3');
cy.get('button').click();
cy.get('p').should('have.text', '-8');
});
```

The `it()` function is used to define an individual test case. The first argument is a string describing what the test case is supposed to do. In this case, it is `'adds two positive numbers correctly'`.

The `cy.visit()` command is used to visit a URL. Here, it's the URL that leads to our calculator's user interface.

The following line is used to enter `'-5'` in the calculator's first input field. The `cy.get()` function is used to obtain elements from the DOM. The `first()` function is used to obtain the first element of the set of corresponding elements. The `type()` command is used to type into a text input field.

Then, select the `'+'` operator from the drop-down menu on our user interface and check that the selected value is indeed `'+'`. The `select()` command is used to select an option from a drop-down menu. The `should()` function is used to make statements about the state of the application.

The following line is used to enter `'-3'` in the calculator's first input field. The `cy.get()` function is used to obtain DOM elements. The `last()` function is used to obtain the last element of the set of corresponding elements. The `type()` command is used to type into a text input field.

Then click on the button to perform the calculation. The `click()` command is used to simulate a mouse click.

Finally, the last line of code checks that the result of the calculation (`'-8'`) is displayed correctly. The `should()` function is again used to make this assertion.

And in our browser, we have the following result:

Figure 8.4 – The "adds two negative numbers correctly" E2E test succeeded

In the next section, we'll learn how to correctly add one positive number and one negative number.

Adds one positive number and one negative number correctly

In this test case, we will see how to write the E2E test to add one positive number and one negative number:

```
it('adds one positive number and one negative number correctly', () =>
{
cy.visit('http://localhost:4200/');
cy.get('input').first().type('5');
cy.get('select').select('+').should('have.value', '+');
cy.get('input').last().type('-3');
cy.get('button').click();
cy.get('p').should('have.text', '2');
});
```

The `it()` function is used to define an individual test case. The first argument is a string describing what the test case is supposed to do. In this case, it is `'adds two positive numbers correctly'`.

The `cy.visit()` command is used to visit a URL. Here, it's the URL that leads to our calculator's user interface.

The following line is used to enter the number `'5'` in the calculator's first input field. The `cy.get()` function is used to obtain elements from the DOM. The `first()` function is used to obtain the first element of the set of corresponding elements. The `type()` command is used to type into a text input field.

Then, select the `'+'` operator from the drop-down menu on our user interface and check that the selected value is indeed `'+'`. The `select()` command is used to select an option from a drop-down menu. The `should()` function is used to make statements about the state of the application.

The following line is used to enter the number `'-3'` in the calculator's first input field. The `cy.get()` function is used to obtain DOM elements. The `last()` function is used to obtain the last element of the set of corresponding elements. The `type()` command is used to type into a text input field.

Then, click on the button to perform the calculation. The `click()` command is used to simulate a mouse click.

Finally, the last line of code checks that the result of the calculation (`'2'`) is displayed correctly. The `should()` function is again used to make this assertion.

And in our browser, we have the following result:

Figure 8.5 – The "adds one positive number and one negative number correctly" E2E test succeeded

In the next section, we'll look at the context of subtraction.

Subtraction context

In this section, we'll look at the various E2E test cases related to subtraction.

Subtracts two positive numbers correctly

In this test case, we will see how to write the E2E test to subtract two positive numbers:

```
it('subtracts two positive numbers correctly', () => {
cy.visit('http://localhost:4200/');
cy.get('input').first().type('5');
cy.get('select').select('-').should('have.value', '-');
cy.get('input').last().type('3');
cy.get('button').click();
cy.get('p').should('have.text', '2');
});
```

The `it()` function is used to define an individual test case. The first argument is a string describing what the test case is supposed to do. In this case, it is `'subtracts two positive numbers correctly'`.

The `cy.visit()` command is used to visit a URL. Here, it's the URL that leads to our calculator's user interface.

The following line is used to enter the number `'5'` in the calculator's first input field. The `cy.get()` function is used to obtain elements from the DOM. The `first()` function is used to obtain the first element of the set of corresponding elements. The `type()` command is used to type into a text input field.

Then, select the `'-'` operator from the drop-down menu on our user interface and check that the selected value is indeed `'-'`. The `select()` command is used to select an option from a drop-down menu. The `should()` function is used to make statements about the state of the application.

The following line is used to enter the number `'3'` in the calculator's first input field. The `cy.get()` function is used to obtain DOM elements. The `last()` function is used to obtain the last element of the set of corresponding elements. The `type()` command is used to type into a text input field.

Then, click on the button to perform the calculation. The `click()` command is used to simulate a mouse click.

Finally, the last line of code checks that the result of the calculation (`'2'`) is displayed correctly. The `should()` function is again used to make this assertion.

And in our browser, we have this result:

Figure 8.6 – The "subtracts two positive numbers correctly" E2E test succeeded

In the next section, we'll learn how to correctly subtract two negative numbers.

Subtracts two negative numbers correctly

In this test case, we will see how to write the E2E test to subtract two negative numbers:

```
it('subtracts two negative numbers correctly', () => {
cy.visit('http://localhost:4200/');
cy.get('input').first().type('-5');
cy.get('select').select('-').should('have.value', '-');
cy.get('input').last().type('-3');
cy.get('button').click();
cy.get('p').should('have.text', '-2');
});
```

In the preceding code, the `it()` function is used to define an individual test case. The first argument is a string describing what the test case is supposed to do. In this case, it is `'subtracts two negative numbers correctly'`.

The `cy.visit()` command is used to visit a URL. Here, it's the URL that leads to our calculator's user interface.

The following line is used to enter the number `'-5'` in the calculator's first input field. The `cy.get()` function is used to obtain elements from the DOM. The `first()` function is used to obtain the first element of the set of corresponding elements. The `type()` command is used to type into a text input field.

Then, select the `'-'` operator from the drop-down menu on our user interface and check that the selected value is indeed `'-'`. The `select()` command is used to select an option from a drop-down menu. The `should()` function is used to make statements about the state of the application.

The following line is used to enter the number `'-3'` in the calculator's first input field. The `cy.get()` function is used to obtain DOM elements. The `last()` function is used to obtain the last element of the set of corresponding elements. The `type()` command is used to type into a text input field.

Then, click on the button to perform the calculation. The `click()` command is used to simulate a mouse click.

Finally, the last line of code checks that the result of the calculation (`'-2'`) is displayed correctly. The `should()` function is again used to make this assertion.

And in our browser, we have this result:

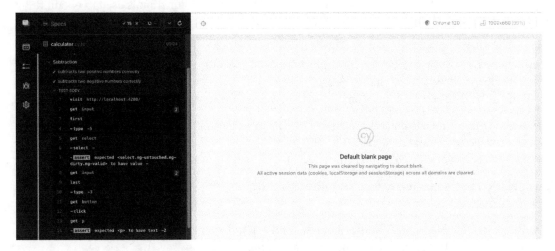

Figure 8.7 – The "subtracts two negative numbers correctly" E2E test succeeded

In the next section, we'll learn how to correctly subtract one positive number and one negative number.

Subtracts one positive number and one negative number correctly

In this test case, we will see how to write the E2E test to subtract one positive number and one negative number:

```
it('subtracts one positive number and one negative number correctly',
() => {
cy.visit('http://localhost:4200/');
cy.get('input').first().type('5');
cy.get('select').select('-').should('have.value', '-');
cy.get('input').last().type('-3');
cy.get('button').click();
cy.get('p').should('have.text', '8');
});
```

The `it()` function is used to define an individual test case. The first argument is a string describing what the test case is supposed to do. In this case, it is `'subtracts one positive number and one negative number correctly'`.

The `cy.visit()` command is used to visit a URL. Here, it's the URL that leads to our calculator's user interface.

The following line is used to enter the number `'5'` in the calculator's first input field. The `cy.get()` function is used to obtain elements from the DOM. The `first()` function is used to obtain the first element of the set of corresponding elements. The `type()` command is used to type into a text input field.

Then, select the `'-'` operator from the drop-down menu on our user interface and check that the selected value is indeed `'-'`. The `select()` command is used to select an option from a drop-down menu. The `should()` function is used to make statements about the state of the application.

The following line is used to enter the number `'-3'` in the calculator's first input field. The `cy.get()` function is used to obtain DOM elements. The `last()` function is used to obtain the last element of the set of corresponding elements. The `type()` command is used to type into a text input field.

Then, click on the button to perform the calculation. The `click()` command is used to simulate a mouse click.

Finally, the last line of code checks that the result of the calculation (`'8'`) is displayed correctly. The `should()` function is again used to make this assertion.

And in our browser, we have this result:

Figure 8.8 – The "subtracts two negative numbers correctly" E2E test succeeded

In the next section, we'll look at the context of multiplication.

Multiplication context

In this section, we'll look at the various E2E test cases related to multiplication.

Multiplies non-zero number and zero correctly

In this test case, we will see how to write the E2E test to multiply a non-zero number and zero:

```
it('multiplies non-zero number and zero correctly', () => {
cy.visit('http://localhost:4200/');
cy.get('input').first().type('5');
cy.get('select').select('*').should('have.value', '*');
cy.get('input').last().type('0');
cy.get('button').click();
cy.get('p').should('have.text', '0');
});
```

The `it()` function is used to define an individual test case. The first argument is a string describing what the test case is supposed to do. In this case, it is `'multiplies non-zero number and zero correctly'`.

The `cy.visit()` command is used to visit a URL. Here, it's the URL that leads to our calculator's user interface.

The following line is used to enter the number `'5'` in the calculator's first input field. The `cy.get()` function is used to obtain elements from the DOM. The `first()` function is used to obtain the first element of the set of corresponding elements. The `type()` command is used to type into a text input field.

Then, select the `'*'` operator from the drop-down menu on our user interface and check that the selected value is indeed `'*'`. The `select()` command is used to select an option from a drop-down menu. The `should()` function is used to make statements about the state of the application.

The following line is used to enter the number `'0'` in the calculator's first input field. The `cy.get()` function is used to obtain DOM elements. The `last()` function is used to obtain the last element of the set of corresponding elements. The `type()` command is used to type into a text input field.

Then, click on the button to perform the calculation. The `click()` command is used to simulate a mouse click.

Finally, the last line of code checks that the result of the calculation (`'0'`) is displayed correctly. The `should()` function is again used to make this assertion.

And in our browser, we have this result:

Figure 8.9 – The "multiplies non-zero number and zero correctly" E2E test succeeded

In the next section, we'll learn how to correctly multiply two positive numbers.

Multiplies two positive numbers correctly

In this test case, we will see how to write the E2E test to multiply two positive numbers:

```
it('multiplies two positive numbers correctly', () => {
cy.visit('http://localhost:4200/');
cy.get('input').first().type('5');
cy.get('select').select('*').should('have.value', '*');
cy.get('input').last().type('3');
cy.get('button').click();
cy.get('p').should('have.text', '15');
});
```

The `it()` function is used to define an individual test case. The first argument is a string describing what the test case is supposed to do. In this case, it is `'multiplies two positive numbers correctly'`.

The `cy.visit()` command is used to visit a URL. Here, it's the URL that leads to our calculator's user interface.

The following line is used to enter the number `'5'` in the calculator's first input field. The `cy.get()` function is used to obtain elements from the DOM. The `first()` function is used to obtain the first element of the set of corresponding elements. The `type()` command is used to type into a text input field.

Then, select the `'*'` operator from the drop-down menu on our user interface and check that the selected value is indeed `'*'`. The `select()` command is used to select an option from a drop-down menu. The `should()` function is used to make statements about the state of the application.

The following line is used to enter the number `'3'` in the calculator's first input field. The `cy.get()` function is used to obtain DOM elements. The `last()` function is used to obtain the last element of the set of corresponding elements. The `type()` command is used to type into a text input field.

Then, click on the button to perform the calculation. The `click()` command is used to simulate a mouse click.

Finally, the last line of code checks that the result of the calculation (`'15'`) is displayed correctly. The `should()` function is again used to make this assertion.

And in our browser, we have this result:

Figure 8.10 – The "multiplies two positive numbers correctly" E2E test succeeded

In the next section, we'll learn how to correctly multiply two negative numbers.

Multiplies two negative numbers correctly

In this test case, we will see how to write the E2E test to multiply two negative numbers:

```
it('multiplies two negative numbers correctly', () => {
cy.visit('http://localhost:4200/');
cy.get('input').first().type('-5');
cy.get('select').select('*').should('have.value', '*');
cy.get('input').last().type('-3');
```

```
cy.get('button').click();
cy.get('p').should('have.text', '15');
});
```

The it() function is used to define an individual test case. The first argument is a string describing what the test case is supposed to do. In this case, it is 'multiplies two negative numbers correctly'.

The cy.visit() command is used to visit a URL. Here, it's the URL that leads to our calculator's user interface.

The following line is used to enter the number '-5' in the calculator's first input field. The cy.get() function is used to obtain elements from the DOM. The first() function is used to obtain the first element of the set of corresponding elements. The type() command is used to type into a text input field.

Then, select the '*' operator from the drop-down menu on our user interface and check that the selected value is indeed '*'. The select() command is used to select an option from a drop-down menu. The should() function is used to make statements about the state of the application.

The following line is used to enter the number '-3' in the calculator's first input field. The cy.get() function is used to obtain DOM elements. The last() function is used to obtain the last element of the set of corresponding elements. The type() command is used to type into a text input field.

Then, click on the button to perform the calculation. The click() command is used to simulate a mouse click.

Finally, the last line of code checks that the result of the calculation ('15') is displayed correctly. The should() function is again used to make this assertion.

And in our browser, we have this result:

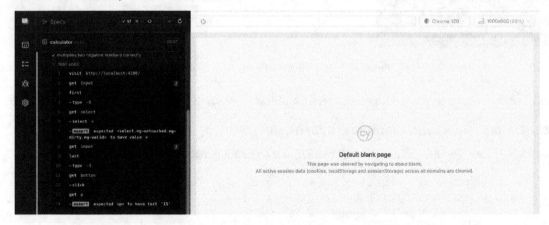

Figure 8.11 – The "multiplies two negative numbers correctly" E2E test succeeded

In the next section, we'll learn how to correctly multiply one positive number and one negative number.

Multiplies one positive number and one negative number correctly

In this test case, we will see how to write the E2E test to multiply one positive number and one negative number:

```
it('multiplies one positive number and one negative number correctly',
() => {
cy.visit('http://localhost:4200/');
cy.get('input').first().type('5');
cy.get('select').select('*').should('have.value', '*');
cy.get('input').last().type('-3');
cy.get('button').click();
cy.get('p').should('have.text', '-15');
});
```

The `it()` function is used to define an individual test case. The first argument is a string describing what the test case is supposed to do. In this case, it is `'multiplies one positive number and one negative number correctly'`.

The `cy.visit()` command is used to visit a URL. Here, it's the URL that leads to our calculator's user interface.

The following line is used to enter the number `'5'` in the calculator's first input field. The `cy.get()` function is used to obtain elements from the DOM. The `first()` function is used to obtain the first element of the set of corresponding elements. The `type()` command is used to type into a text input field.

Then, select the `'*'` operator from the drop-down menu on our user interface and check that the selected value is indeed `'*'`. The `select()` command is used to select an option from a drop-down menu. The `should()` function is used to make statements about the state of the application.

The following line is used to enter the number `'-3'` in the calculator's first input field. The `cy.get()` function is used to obtain DOM elements. The `last()` function is used to obtain the last element of the set of corresponding elements. The `type()` command is used to type into a text input field.

Then, click on the button to perform the calculation. The `click()` command is used to simulate a mouse click.

Finally, the last line of code checks that the result of the calculation (`'-15'`) is displayed correctly. The `should()` function is again used to make this assertion.

And in our browser, we have this result:

Figure 8.12 – The "multiplies one positive number and one negative number correctly" E2E test succeeded

In the next section, we'll look at the context of division.

Division context

In this section, we'll look at the various E2E test cases related to division.

Divides a positive non-zero number by another positive non-zero number

In this test case, we will see how to write the E2E test to divide a positive non-zero number by another positive non-zero number:

```
it('divides a positive non-zero number by another positive non-zero
number', () => {
    cy.visit('http://localhost:4200/');
    cy.get('input').first().type('5');
    cy.get('select').select('/').should('have.value', '/');
    cy.get('input').last().type('2');
    cy.get('button').click();
    cy.get('p').should('have.text', '2.5');
});
```

The it() function is used to define an individual test case. The first argument is a string describing what the test case is supposed to do. In this case, it is 'divides a positive non-zero number by another positive non-zero number'.

The cy.visit() command is used to visit a URL. Here, it's the URL that leads to our calculator's user interface.

The following line is used to enter the number '5' in the calculator's first input field. The cy.get() function is used to obtain elements from the DOM. The first() function is used to obtain the first element of the set of corresponding elements. The type() command is used to type into a text input field.

Then, select the '/' operator from the drop-down menu on our user interface and check that the selected value is indeed '/'. The select() command is used to select an option from a drop-down menu. The should() function is used to make statements about the state of the application.

The following line is used to enter the number '2' in the calculator's first input field. The cy.get() function is used to obtain DOM elements. The last() function is used to obtain the last element of the set of corresponding elements. The type() command is used to type into a text input field.

Then, click on the button to perform the calculation. The click() command is used to simulate a mouse click.

Finally, the last line of code checks that the result of the calculation ('2.5') is displayed correctly. The should() function is again used to make this assertion.

And in our browser, we have this result:

Figure 8.13 – The "divides a positive non-zero number by another
positive non-zero number" E2E test succeeded

In the next section, we'll look at the test case for dividing a non-zero negative number by a non-zero positive number.

Divides a negative non-zero number by a positive non-zero number

In this test case, we will see how to write the E2E test to divide a negative non-zero number by another positive non-zero number:

```
it('divides a negative non-zero number by another positive non-zero
number', () => {
    cy.visit('http://localhost:4200/');
    cy.get('input').first().type('-5');
    cy.get('select').select('/').should('have.value', '/');
    cy.get('input').last().type('2');
    cy.get('button').click();
    cy.get('p').should('have.text', '-2.5');
});
```

The `it()` function is used to define an individual test case. The first argument is a string describing what the test case is supposed to do. In this case, it is `'divides a negative non-zero number by another positive non-zero number'`.

The `cy.visit()` command is used to visit a URL. Here, it's the URL that leads to our calculator's user interface.

The following line is used to enter the number `'-5'` in the calculator's first input field. The `cy.get()` function is used to obtain elements from the DOM. The `first()` function is used to obtain the first element of the set of corresponding elements. The `type()` command is used to type into a text input field.

Then, select the `'/'` operator from the drop-down menu on our user interface and check that the selected value is indeed `'/'`. The `select()` command is used to select an option from a drop-down menu. The `should()` function is used to make statements about the state of the application.

The following line is used to enter the number `'2'` in the calculator's first input field. The `cy.get()` function is used to obtain DOM elements. The `last()` function is used to obtain the last element of the set of corresponding elements. The `type()` command is used to type into a text input field.

Then, click on the button to perform the calculation. The `click()` command is used to simulate a mouse click.

Finally, the last line of code checks that the result of the calculation (`'-2.5'`) is displayed correctly. The `should()` function is again used to make this assertion.

And in our browser, we have this result:

Figure 8.14 – The "divides a negative non-zero number by another
positive non-zero number" E2E test succeeded

In the next section, we'll look at the test case for dividing a non-zero negative number by another non-zero negative number.

Divides a negative non-zero number by another negative non-zero number

In this test case, we will see how to write the E2E test to divide a negative non-zero number by another negative non-zero number:

```
it('divides a negative non-zero number by another negative non-zero
number', () => {
    cy.visit('http://localhost:4200/');
    cy.get('input').first().type('-5');
    cy.get('select').select('/').should('have.value', '/');
    cy.get('input').last().type('-2');
    cy.get('button').click();
    cy.get('p').should('have.text', '2.5');
});
```

The it() function is used to define an individual test case. The first argument is a string describing what the test case is supposed to do. In this case, it is 'divides a negative non-zero number by another negative non-zero number'.

The cy.visit() command is used to visit a URL. Here, it's the URL that leads to our calculator's user interface.

The following line is used to enter the number '-5' in the calculator's first input field. The cy.get() function is used to obtain elements from the DOM. The first() function is used to obtain the first element of the set of corresponding elements. The type() command is used to type into a text input field.

Then, select the '/' operator from the drop-down menu on our user interface and check that the selected value is indeed '/'. The select() command is used to select an option from a drop-down menu. The should() function is used to make statements about the state of the application.

The following line is used to enter the number '-2' in the calculator's first input field. The cy.get() function is used to obtain DOM elements. The last() function is used to obtain the last element of the set of corresponding elements. The type() command is used to type into a text input field.

Then, click on the button to perform the calculation. The click() command is used to simulate a mouse click.

Finally, the last line of code checks that the result of the calculation ('2.5') is displayed correctly. The should() function is again used to make this assertion.

And in our browser, we have this result:

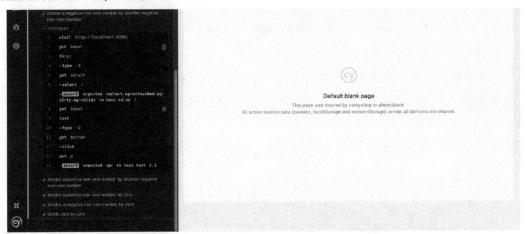

Figure 8.15 – The "divides a negative non-zero number by another
negative non-zero number" E2E test succeeded

In the next section, we'll look at the test case for dividing a non-zero positive number by a non-zero negative number.

Divides a positive non-zero number by a negative non-zero number

In this test case, we will see how to write the E2E test to divide a positive non-zero number by another negative non-zero number:

```
it('divides a positive non-zero number by another negative non-zero
number', () => {
    cy.visit('http://localhost:4200/');
    cy.get('input').first().type('5');
    cy.get('select').select('/').should('have.value', '/');
    cy.get('input').last().type('-2');
    cy.get('button').click();
    cy.get('p').should('have.text', '-2.5');
});
```

The `it()` function is used to define an individual test case. The first argument is a string describing what the test case is supposed to do. In this case, it is `'divides a positive non-zero number by another negative non-zero number'`.

The `cy.visit()` command is used to visit a URL. Here, it's the URL that leads to our calculator's user interface.

The following line is used to enter the number `'5'` in the calculator's first input field. The `cy.get()` function is used to obtain elements from the DOM. The `first()` function is used to obtain the first element of the set of corresponding elements. The `type()` command is used to type into a text input field.

Then, select the `'/'` operator from the *drop-down menu* on our user interface and check that the selected value is indeed `'/'`. The `select()` command is used to select an option from a drop-down menu. The `should()` function is used to make statements about the state of the application.

The following line is used to enter the number `'-2'` in the calculator's first input field. The `cy.get()` function is used to obtain DOM elements. The `last()` function is used to obtain the last element of the set of corresponding elements. The `type()` command is used to type into a text input field.

Then, click on the button to perform the calculation. The `click()` command is used to simulate a mouse click.

Finally, the last line of code checks that the result of the calculation (`'-2.5'`) is displayed correctly. The `should()` function is again used to make this assertion.

And in our browser, we have this result:

Figure 8.16 – The "divides a positive non-zero number by another
negative non-zero number" E2E test succeeded

In the next section, we'll look at the test case for dividing a non-zero positive number by zero.

Divides a positive non-zero number by zero

In this test case, we will see how to write the E2E test to divide a positive non-zero number by zero:

```
it('divides a positive non-zero number by zero', () => {
    cy.visit('http://localhost:4200/');
    cy.get('input').first().type('5');
    cy.get('select').select('/').should('have.value', '/');
    cy.get('input').last().type('0');
    cy.get('button').click();
    cy.get('p').should('have.text', 'Infinity');
});
```

The `it()` function is used to define an individual test case. The first argument is a string describing what the test case is supposed to do. In this case, it is `'divides a positive non-zero number by zero'`.

The `cy.visit()` command is used to visit a URL. Here, it's the URL that leads to our calculator's user interface.

The following line is used to enter the number `'5'` in the calculator's first input field. The `cy.get()` function is used to obtain elements from the DOM. The `first()` function is used to obtain the first element of the set of corresponding elements. The `type()` command is used to type into a text input field.

Then, select the '/' operator from the drop-down menu on our user interface and check that the selected value is indeed '/'. The select() command is used to select an option from a drop-down menu. The should() function is used to make statements about the state of the application.

The following line is used to enter the number '0' in the calculator's first input field. The cy.get() function is used to obtain DOM elements. The last() function is used to obtain the last element of the set of corresponding elements. The type() command is used to type into a text input field.

Then, click on the button to perform the calculation. The click() command is used to simulate a mouse click.

Finally, the last line of code checks that the result of the calculation ('Infinity') is displayed correctly. The should() function is again used to make this assertion.

And in our browser, we have this result:

Figure 8.17 – The "divides a positive non-zero number by zero" E2E test succeeded

In the next section, we'll look at the test case for dividing a non-zero negative number by zero.

Divides a negative non-zero number by zero

In this test case, we will see how to write the E2E test to divide a negative non-zero number by zero:

```
it('divides a negative non-zero number by zero', () => {
    cy.visit('http://localhost:4200/');
    cy.get('input').first().type('-5');
    cy.get('select').select('/').should('have.value', '/');
    cy.get('input').last().type('0');
    cy.get('button').click();
    cy.get('p').should('have.text', '-Infinity');
});
```

The `it()` function is used to define an individual test case. The first argument is a string describing what the test case is supposed to do. In this case, it is `'divides a negative non-zero number by zero'`.

The `cy.visit()` command is used to visit a URL. Here, it's the URL that leads to our calculator's user interface.

The following line is used to enter the number `'-5'` in the calculator's first input field. The `cy.get()` function is used to obtain elements from the DOM. The `first()` function is used to obtain the first element of the set of corresponding elements. The `type()` command is used to type into a text input field.

Then, select the `'/'` operator from the drop-down menu on our user interface and check that the selected value is indeed `'/'`. The `select()` command is used to select an option from a drop-down menu. The `should()` function is used to make statements about the state of the application.

The following line is used to enter the number `'0'` in the calculator's first input field. The `cy.get()` function is used to obtain DOM elements. The `last()` function is used to obtain the last element of the set of corresponding elements. The `type()` command is used to type into a text input field.

Then, click on the button to perform the calculation. The `click()` command is used to simulate a mouse click.

Finally, the last line of code checks that the result of the calculation (`'-Infinity'`) is displayed correctly. The `should()` function is again used to make this assertion.

And in our browser, we have this result:

Figure 8.18 - Divide a negative non-zero number by zero end-to-end test succeeded

In the next section, we'll look at the test case for dividing a non-zero negative number by zero.

Divides zero by zero

In this test case, we will see how to write the E2E test to divide zero by zero:

```
it('divide zero by zero', () => {
    cy.visit('http://localhost:4200/');
    cy.get('input').first().type('0');
    cy.get('select').select('/').should('have.value', '/');
    cy.get('input').last().type('0');
    cy.get('button').click();
    cy.get('p').should('have.text', 'NaN');
});
```

The `it()` function is used to define an individual test case. The first argument is a string describing what the test case is supposed to do. In this case, it is `'divide zero by zero'`.

The `cy.visit()` command is used to visit a URL. Here, it's the URL that leads to our calculator's user interface.

The following line is used to enter the number `'0'` in the calculator's first input field. The `cy.get()` function is used to obtain elements from the DOM. The `first()` function is used to obtain the first element of the set of corresponding elements. The `type()` command is used to type into a text input field.

Then, select the `'/'` operator from the drop-down menu on our user interface and check that the selected value is indeed `'/'`. The `select()` command is used to select an option from a drop-down menu. The `should()` function is used to make statements about the state of the application.

The following line is used to enter the number `'0'` in the calculator's first input field. The `cy.get()` function is used to obtain DOM elements. The `last()` function is used to obtain the last element of the set of corresponding elements. The `type()` command is used to type into a text input field.

Then, click on the button to perform the calculation. The `click()` command is used to simulate a mouse click.

Finally, the last line of code checks that the result of the calculation (`'NaN'`) is displayed correctly. The `should()` function is again used to make this assertion.

And in our browser, we have this result:

Figure 8.19 – The "divide zero by zero" E2E test succeeded

In the next section, we'll look at what Cypress custom commands are and how they can be used to make code easier to maintain and read.

Using Cypress custom commands

Cypress custom commands are user-defined actions and assertions that extend the capabilities of the Cypress test framework. They enable testers to encapsulate repetitive actions, streamline automation workflows, and optimize test scripts for specific needs. Custom commands can be added or replaced, offering a flexible way of interacting with web applications and improving the efficiency and readability of test scripts.

To create a custom command in Cypress, you use the `Cypress.Commands.add()` method. This method lets you define a new command that can be used throughout your test suite.

For example, we could create a custom command to test the `add()` operation. To achieve this, we'll add some code to the `commands.ts` file in the `support` folder contained in the `cypress` folder:

```
Cypress.Commands.add(
  'performCalculation',
  (firstNumber, operator, secondNumber) => {
    cy.visit('http://localhost:4200/');
```

```
      cy.get('input').first().type(firstNumber);
      cy.get('select').select(operator).should('have.value', operator);
      cy.get('input').last().type(secondNumber);
      cy.get('button').click();
    }
);

declare namespace Cypress {
  interface Chainable<Subject = any> {
    performCalculation(
      firstNumber: string,
      operator: string,
      secondNumber: string
    ): Chainable<any>;
  }
}
```

Here's a breakdown of the code:

- `Cypress.Commands.add('performCalculation', (firstNumber, operator, secondNumber) => {...})`: This defines a new custom command named `performCalculation`. This command takes three parameters: `firstNumber`, `operator`, and `secondNumber`.

- `cy.get('input').first().type(firstNumber)`: This selects the first input element on the page and enters the `firstNumber` value.

- `cy.get('select').select(operator).should('have.value', operator)`: This selects a drop-down list (or selection element) and chooses the option corresponding to the operator argument. It then asserts that the selected value corresponds to the operator.

- `cy.get('input').last().type(secondNumber)`: This selects the last input element on the page and types in the `secondNumber` value.

- `cy.get('button').click()`: This clicks on a button, presumably to perform the calculation.

Now, at the level of our `calculator.cy.ts` file, we will do the following to set up our E2E test:

```
it('adds two positive numbers correctly', () => {
  cy.performCalculation('5', '+', '3');
  cy.get('p').should('have.text', '8');
});
```

Here's a breakdown of the code:

- `cy.performCalculation('5', '+', '3')`: A custom command that presumably performs a calculation operation within the application being tested. It takes three arguments: the first number, the operator, and the second number. In this case, it's adding 5 and 3.

- `cy.get('p').should('have.text', '8')`: An assertion that checks whether the selected element has the specified text. Here, it's checking whether the paragraph element contains the text `'8'`.

Now, in our browser, we have this result:

Figure 8.20 – The "adds two positive numbers correctly" E2E test succeeded using Cypress.Command

Let's summarize the chapter now.

Summary

This chapter covered the essential aspects of structuring, writing, and improving E2E tests using Cypress, a popular testing framework for web applications. It began by explaining the importance of structuring E2E tests to ensure they are complete, maintainable, and easy to understand. This involves organizing tests into logical groups using `describe` and `it` blocks, which help categorize tests according to the functionality or feature they test.

Writing E2E test cases was a crucial aspect covered in this chapter. It focused on the use of Cypress commands and assertions to interact with web elements and verify their behavior. The chapter provided examples of how to use commands such as `cy.visit()`, `cy.get()`, `cy.contains()`, `cy.type()`, and `cy.should()` to navigate the application, interact with elements, and validate the application state. It stressed the importance of structuring tests in such a way as to move from request to command or assertion, ensuring that tests are deterministic and reliable.

This chapter also explored the use of Cypress custom commands to improve the readability and maintainability of E2E tests. Custom commands enable developers to encapsulate repetitive patterns and actions in reusable functions, making tests more readable and easier to understand. This is achieved by defining custom commands using `Cypress.Commands.add()`, which can then be used in place of standard Cypress commands in tests. The chapter provided examples of how to create custom commands for actions such as selecting form elements by their labels, which simplifies the test code and makes it more consistent with the actions a real user would perform.

In summary, this chapter provides a comprehensive guide to structuring, writing, and improving E2E tests with Cypress, emphasizing the importance of clear test organization, effective use of Cypress commands and assertions, and the creation of custom commands to improve test readability and maintainability.

In the next chapter, we'll learn about **Continuous Integration (CI)** and **Continuous Deployment (CD)**.

Part 4:
Continuous Integration and Continuous Deployment for Angular Applications

In this section, you'll start by learning about the general principles of **continuous integration** (**CI**) and **continuous deployment** (**CD**). Next, you'll embark on an in-depth exploration of TDD best practices and patterns. Finally, you'll learn about refactoring and code improvement.

This part has the following chapters:

- *Chapter 9, Understanding Continuous Integration and Continuous Deployment (CI/CD)*
- *Chapter 10, Best Practices and Patterns for Angular TDD*
- *Chapter 11, Refactoring and Improving Angular Code through TDD*

9

Understanding Continuous Integration and Continuous Deployment (CI/CD)

In software development, the combination of **continuous integration and deployment** (**CI/CD**) practices with **test-driven development** (**TDD**) has proven to be a powerful combination for delivering high-quality software at a rapid pace. **Continuous integration** (**CI**) and TDD work in synergy to automate the software development cycle, promote a culture of testing, and enable teams to build and deploy code iteratively and with confidence.

CI/CD covers a series of automated steps that streamline the software delivery pipeline, from code changes made by developers to the deployment of these changes in production. By implementing CI/CD practices, development teams can reduce manual errors, improve collaboration between team members, increase delivery speed, and improve overall software quality.

In this chapter, we'll explore the fundamental concepts of CI/CD, explore the benefits of adopting these practices, discuss the key components of a CI/CD pipeline, and provide an overview of how organizations can successfully implement CI/CD processes to streamline their **software development life cycle** (**SDLC**).

In summary, here are the main topics that will be covered in this chapter:

- Understanding continuous integration and continuous deployment
- Setting up CI/CD pipelines for automating build with GitHub Actions
- Setting up CI/CD pipelines for automating tests with GitHub Actions
- Setting up CI/CD pipelines for automating deployment processes with GitHub Actions

Technical requirements

To follow along with the examples and exercises in this chapter, you will need to have a basic understanding of Angular and TypeScript, as well as the following technical requirements:

- Node.js and npm installed on your computer
- Angular CLI installed globally
- A code editor, such as Visual Studio Code, installed on your computer

The code files required for this chapter can be found on GitHub at `https://github.com/PacktPublishing/Mastering-Angular-Test-Driven-Development/tree/main/Chapter%209`.

Understanding CI and CD

CI and CD have become essential practices in modern software development, enabling teams to deliver high-quality code quickly and efficiently. This comprehensive guide covers the fundamental concepts of CI/CD, explores their benefits, discusses best practices, and outlines successful implementation strategies. From understanding the fundamentals to optimizing CI/CD pipelines, this chapter aims to equip readers with the knowledge and tools they need to harness the power of CI/CD and transform software delivery.

But before we go ahead and use these practices, let's understand them.

What is CI?

CI is a DevOps software development practice in which developers regularly merge their code changes into a central repository. After each merge, automated builds and tests are run to ensure that the new code is error free and meets the project's quality standards. This process is essential for identifying and correcting bugs more quickly, improving software quality, and reducing the time needed to validate and release new software updates. CI encourages frequent code integration, often several times a day, so that integration problems can be identified early in the development cycle and corrected more easily. It also encourages a cultural shift towards more frequent code deliveries, which is essential if we are to reap the benefits of CI's automation and efficiency.

CI is the first phase of the CI/CD pipeline, part of the broader DevOps approach to software development. It follows the agile software development methodology, where work is broken down into small, manageable tasks that can be completed and integrated frequently. The use of CI tools, such as GitHub Actions, Jenkins, Buildbot, Go, Travis CI, and GitLab CI, facilitates the automation of build and test processes, making it easier for developers to integrate their changes with the rest of the project and identify issues early in the development process.

The benefits of CI include improved developer productivity, faster delivery of updates, and a more predictable delivery schedule. It also improves cross-team collaboration and systems integration, reducing testing errors and improving the efficiency of the software development cycle. However, the challenges of CI mainly concern team adoption and the initial technical installation of CI tools. Overcoming these challenges and effectively implementing CI practices is essential to realizing the full potential of CI in improving software development processes and outcomes.

In the next section, we'll explore in more depth the benefits of CI for development teams.

Benefits of CI for development teams

The benefits of CI for development teams are manifold, encompassing efficiency, quality, and customer satisfaction. Here's a detailed overview of these benefits:

- **Faster iteration and problem resolution**: CI enables teams to integrate code changes more frequently, speeding up iterations and facilitating problem resolution. Small code changes are simpler to manage, reducing the complexity of problems that can arise.

- **Improved code quality and fewer bugs**: By frequently integrating and testing code, CI enables bugs to be identified and corrected early in the development cycle. The result is higher-quality code with fewer defects, improving the user experience and reducing downtime.

- **Increased efficiency and reduced costs**: CI-driven automation reduces manual tasks, saving time for developers. This not only increases efficiency but also reduces the costs associated with manual testing and error management. As a result, engineers can devote more time to value-added activities.

- **Improved transparency and collaboration**: CI promotes transparency by providing continuous feedback on code quality and integration issues. It also promotes better team collaboration by ensuring that code changes are integrated and tested regularly, enabling better coordination between team members.

- **Faster time to market**: By automating the build, test, and deployment processes, CI enables teams to deliver new features and updates to end users more quickly. This responsiveness keeps the development team competitive and ensures that customers benefit from the latest enhancements.

- **Improved customer satisfaction**: Fewer bugs and errors end up in production, improving the user experience. CI also enables rapid response to customer feedback, enabling teams to make adjustments and improvements more efficiently.

- **Reduced mean time to resolution (MTTR)**: CI enables problems to be detected and resolved more quickly, thus reducing the MTTR. This ensures that the software remains stable and reliable, minimizing downtime.

- **Increased test reliability**: Continuous testing within the CI framework improves test reliability by enabling more precise tests to be carried out. This ensures that the software is thoroughly tested and ready for production, boosting confidence in software quality.

- **Competitive advantage**: Organizations that adopt **business intelligence (BI)** have a competitive advantage because they can deploy functionality more quickly, which in turn saves them money. This early feedback and automation helps to reduce lead times, deployment frequency, and change failure rates, which in turn improves business results.

- **Increased transparency and accountability within the team**: CI/CD practices increase transparency and accountability within the team, enabling problems to be identified and resolved quickly, including construction failures and architectural setbacks. This continuous feedback loop improves overall product quality.

In summary, CI offers development teams significant benefits in terms of efficiency, quality, and customer satisfaction. It streamlines the development process, reduces costs, and enhances collaboration, ultimately leading to the delivery of high-quality software products more quickly and reliably. In the next section, we will explore the key principles of CI implementation.

Key principles of CI implementation

The key principles of CI implementation aim to improve the efficiency, quality, and speed of software development. Here are the key principles:

- **Automate everything**: CI focuses on automating the build, test, and integration processes. Automation reduces manual effort, minimizes errors, and accelerates the development cycle.

- **Frequent integration**: Frequently integrate code changes into a shared repository, ideally several times a day. This practice enables integration problems to be identified and resolved early in the development cycle.

- **Make the build process fast**: The build process should be as fast as possible to ensure rapid feedback. Rapid construction means that problems can be detected and resolved more quickly, facilitating continuous improvement.

- **Immediate feedback**: CI relies on immediate feedback from automated builds and tests. This feedback is essential for identifying and resolving problems early in the development process.

- **Start small and grow**: Start with a simple CI configuration and gradually add other tools and practices as required. This approach encourages flexibility and experimentation, allowing teams to find what works best in their specific context.

- **Define success indicators**: Clearly define success indicators for your CI process, such as accelerated code construction or reduced error and work rates. Use these indicators to measure the effectiveness of your CI practices and to guide improvements.

- **Documentation**: Document the CI process and the tools used by all developers and stakeholders. Good documentation ensures that everyone understands how to contribute to the CI process and solve problems efficiently.

- **Collaboration between operations and development**: Encourage a culture in which operations and development work closely together. This collaboration is essential for understanding software reliability and performance from both points of view.

- **Scalability**: CI breaks down barriers to growth by automating code integration and communication, allowing organizations to scale their development teams, code base, and infrastructure.

- **Investment in the learning curve**: Implementing CI successfully involves learning new skills in areas such as version control and automation. However, these skills are readily available, and the benefits of CI outweigh the initial investment.

These principles guide the implementation of CI, ensuring that it becomes an integral part of the software development process, improving productivity, quality, and speed. In the next section, we will learn what **continuous deployment (CD)** is.

What is CD?

CD is an automated software release practice in which code changes are automatically deployed at different stages as they pass predefined tests. The aim of CD is to accelerate production releases by using automation to minimize human intervention during the deployment process. This approach is part of the wider DevOps practice, which aims to accelerate innovation and value creation by applying automation to every stage of the SDLC.

Software design requires a mix of rigorous testing, close collaboration between teams, advanced tools, and workflow processes throughout the application design and development process. When successfully implemented, CD enables organizations to respond quickly to customer requests and deliver software updates rapidly, often within minutes of validating code changes. This process includes automating build, test, and deployment in a single workflow, with the aim of automating software deployment in production.

The benefits of CD include fully automated deployment cycles, enabling organizations to spend more time on software creation than on release preparation. It also leads to more regular, incremental deployments, facilitating faster product development and a continuous improvement model. In addition, CD provides rapid feedback loops on new features, updates, and code changes, enabling organizations to quickly receive and integrate user feedback.

CD goes further than CI, which automates everything right up to the deployment itself, requiring human intervention to set up the deployment. CD automates the whole process, including the release of the software itself, making it a natural evolution of CD if the pipeline is properly set up and designed to test all elements of a software product before release.

A CD pipeline streamlines software delivery by automatically building, testing, and deploying code changes directly to production. It involves automated testing and monitoring throughout the pipeline to detect potential errors, functional problems, and bugs, providing real-time alerts and preventing problems from reaching the main software branch or production. This approach underlines the main objective of DevOps: the CD of value to end users.

In practice, this means that a change made by a developer to a cloud application can be put into production within minutes of being written, provided it passes automated testing. This makes it much easier to receive and integrate user feedback on an ongoing basis. However, delivering value depends heavily on well-designed test automation, which can require a significant initial investment.

Overall, CD is an essential aspect of the DevOps approach, enabling organizations to release software updates quickly and efficiently, accelerating innovation and value creation for end-users.

Benefits of CD for development teams

CD offers several key benefits to development teams, facilitating a more efficient, agile, and responsive software development process. Here are the main benefits:

- **Fully automated deployment cycles**: CD enables organizations to automate the entire deployment process, reducing manual intervention and allowing development teams to focus more on coding and less on release preparation. This automation speeds up the deployment of new features and updates, enabling teams to deliver software faster and more efficiently.

- **More regular, incremental deployments**: By automating deployments, CD enables small, incremental changes to be released more frequently. This approach enables faster product development and facilitates a continuous improvement model, in which teams can rapidly iterate on their software based on user feedback and market demands.

- **Rapid feedback loops on new features**: CD provides real-time feedback on new features, updates, and code changes. This immediate feedback loop is essential to enable teams to rapidly adapt and improve their software, ensuring that the final product meets users' expectations and requirements.

- **Event response**: The CD enables teams to react quickly to system errors in production, security incidents or potential new features to be developed during web application development. Immediate release of code to production enables organizations to address and resolve issues more quickly, with metrics such as MTTR enabling response times to be assessed and improved over time.

- **Streamlined release cycles for faster time to market**: By automating the deployment process, CD enables software development teams to quickly deliver new features and bug fixes to end users. This automation reduces the risk of human error and enables small, frequent updates to be deployed quickly, speeding up time to market and giving companies a competitive edge.

- **Early detection of problems thanks to automated testing**: CD emphasizes the importance of automated testing throughout the software development process. By carrying out continuous testing, developers can quickly identify and resolve any potential problems, thus guaranteeing the stability and reliability of the software. This early detection helps reduce the likelihood of costly errors in production and instills confidence in the development team.

- **Continuous feedback loop for continuous improvement**: CD fosters a culture of continuous improvement by establishing a feedback loop between developers and end users. This iterative process enables organizations to adapt and respond to changing user needs, ensuring that their software remains relevant and competitive.

- **Improved collaboration and communication**: CD promotes collaboration and communication between team members, improving the overall efficiency of the development process. By automating the deployment pipeline, developers can concentrate on their core tasks, facilitating seamless integration between different teams and resulting in faster, more efficient software releases.

In short, CD offers development teams the ability to deliver software faster, ensure high quality through automated testing, and maintain a responsive and agile development process. These benefits collectively contribute to a more efficient, innovative, and customer-centric software development cycle.

Key principles of CD implementation

Overall, CD is an essential aspect of the DevOps approach, enabling organizations to release software updates quickly and efficiently, accelerating innovation and value creation for end users. The key principles of CD implementation are essential to creating a streamlined and automated software release process. These principles derive from a combination of agile and organizational best practices aimed at delivering software to end users as quickly as possible, learning from their experience and incorporating their feedback into the next release. Here are the fundamental principles:

- **Build quality**: This principle emphasizes building quality into the product from the outset, rather than relying on inspection to achieve it. It involves creating and evolving feedback loops to detect problems at an early stage, ideally before they are recorded in the version control system. Automated testing should be used to detect defects before they worsen over time.

- **Work in small batches**: CD encourages working with small, manageable changes rather than large, infrequent releases. This approach reduces the time needed to obtain feedback, facilitates problem identification and resolution, and increases efficiency and motivation. The aim is to change the economics of the software delivery process to make it viable to work in small batches.

- **Computers do repetitive tasks, people solve problems**: This principle emphasizes the importance of automating repetitive tasks, such as regression testing, so that humans can concentrate on solving problems. The aim is to create a balance in which computers handle the simple, repetitive tasks, and humans the more complex, creative ones.

- **Continuous improvement**: CD promotes the idea of continuous improvement, or *kaizen*, derived from the Lean movement. It's about seeing improvement work as an essential part of everyday work, and constantly striving to make things better. It's about not being satisfied with the status quo and always looking for opportunities to improve.

- **Everyone is responsible**: In successful organizations, everyone is responsible for the quality and stability of the software they build. This principle encourages a collaborative approach in which developers, operational teams, and other stakeholders work together to achieve the organization's goals, rather than optimizing the success of their own team. It emphasizes the importance of rapid feedback loops based on customer feedback and organizational impact.

Implementing these principles requires a cultural change within the organization, fostering a collaborative environment where everyone is encouraged to ensure that the product delivered to the end user is of the highest possible quality. This means tackling tedious or error-prone tasks early on in the process to avoid aggravating problems and optimize the use of resources. In the next section, we will learn how to set up CI/CD pipelines for automating build with GitHub Actions.

Setting up CI/CD pipelines for automating build with GitHub Actions

Setting up CI/CD pipelines with GitHub Actions involves several steps, each of which is crucial to automating build processes. Here's a step-by-step guide to get you started.

Step 1 – create or choose a repository and project

Start by selecting a repository in which you wish to set up your CI/CD pipeline. This can be an existing project or a new one you're working on. In our case, it will be this repository on GitHub: `https://github.com/PacktPublishing/Mastering-Angular-Test-Driven-Development/tree/main`.

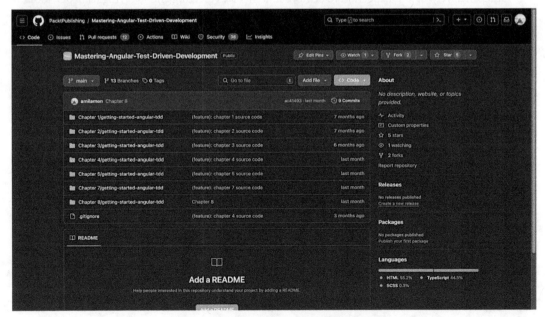

Figure 9.1 – Calculator repository project

Step 2 – open GitHub Actions in your project repository

Now, go to the GitHub Actions tab in the top navigation bar of your repository. Here, you'll find a variety of CI/CD automation templates and workflows tailored to your project's technology stack. GitHub Actions offers a wide range of predefined workflows and lets you create your own from scratch.

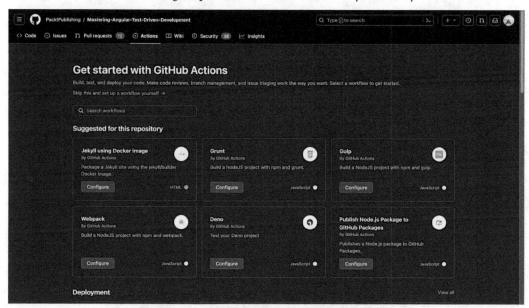

Figure 9.2 – GitHub Actions pipeline templates

Step 3 – define your CI/CD workflow

Our project is an Angular project, so it runs on Node.js. We'll therefore choose the GitHub Actions template dedicated to Node.js, which we'll modify as we go along so that it meets our needs. You need to search using the `node` keyword in the GitHub Actions template search bar, filtering by the **Continuous integration** category:

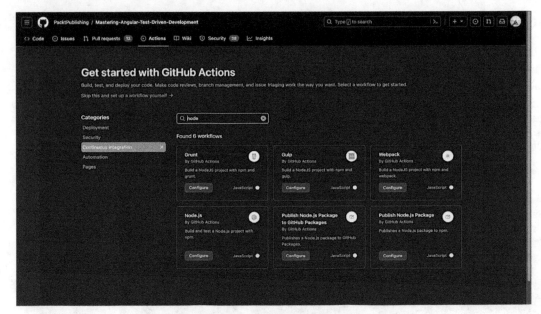

Figure 9.3 – GitHub Actions pipeline templates

As we can see in *Figure 9.4*, Node.js is in the list:

Figure 9.4 – Node.js by GitHub Actions

We can now click on the **Configure** button and we'll be redirected to the interface shown in *Figure 9.5*:

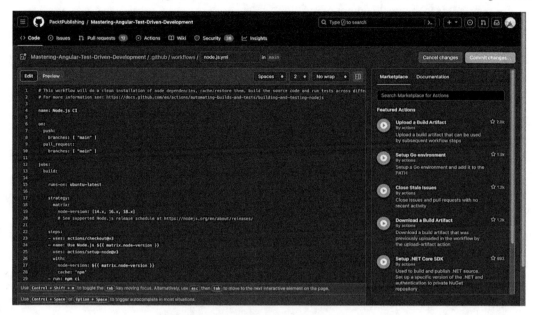

Figure 9.5 – Node.js basic pipeline template

We can now begin the modifications. First, we'll change the name of the file at the top to `angular-tdd.yml`, as shown in *Figure 9.6*:

Figure 9.6 – Workflow name

Next, we can modify the value of name at the beginning of our file. Instead of `Node.js CI`, we'll call it `Angular TDD CI/CD`:

Figure 9.7 – Pipeline name

Next, we can change the `-version: [14.x, 16.x, 18.x]` array node to `node-version: [18.x]`:

```
19        node-version: [18.x]
20        # See supported Node.js release schedule at https://nodejs.org/en/about/releases/
```

Figure 9.8 – Pipeline Node.js version

Finally, we'll delete the last line of our file (i.e., - run: npm test) because we don't have any tests at the moment. This is the final content of our file:

```
# This workflow will do a clean installation of node dependencies,
cache/restore them, build the source code and run tests across
different versions of node
# For more information see: https://docs.github.com/en/actions/
automating-builds-and-tests/building-and-testing-nodejs


name: Angular TDD CI/CD

on:
  push:
    branches: [ "main" ]
  pull_request:
    branches: [ "main" ]

jobs:
  build:

    runs-on: ubuntu-latest

    strategy:
      matrix:
        node-version: [18.x]
        # See supported Node.js release schedule at https://nodejs.
org/en/about/releases/

    steps:
    - uses: actions/checkout@v3
    - name: Use Node.js ${{ matrix.node-version }}
      uses: actions/setup-node@v3
      with:
        node-version: ${{ matrix.node-version }}
        cache: 'npm'
        cache-dependency-path: '**/package-lock.json'
    - run: npm ci
    - run: npm run build --if-present
```

Now, we can save the file by clicking on the **Commit changes** button.

This workflow is designed to automate the process of installing Node.js dependencies, caching them for faster future builds and building the Angular project. The following is a decomposition of the key components of our workflow and their functions:

- **Job definition**: The workflow defines a single job named `build`. This job runs on the latest Ubuntu virtual machine provided by GitHub Actions.

- **Working directory**: The `defaults` section sets the working directory for all stages of the build job to `./Chapter 9/getting-started-angular-tdd/`. This ensures that commands are executed in the correct location where your Angular project is located.

- **Node.js matrix**: The `strategy` section defines a matrix that runs the job multiple times, each time with a different Node.js version. In this example, the matrix includes only one version: `18.x`. You can expand this to include more versions for broader compatibility testing.

- **Checkout**: The first step (`uses: actions/checkout@v3`) uses the official GitHub Actions `checkout` action to clone the repository's code onto the runner.

- **Set up Node.js**: The second step (`uses: actions/setup-node@v3`) uses the official GitHub Actions `setup-node` action to install and configure the specified Node.js version (`18.x`) on the runner.

- The `cache` parameter is set to npm to enable caching of Node.js modules between workflow runs, potentially speeding up subsequent executions. `cache-dependency-path` is set to `**/package-lock.json` to ensure the cache is invalidated if the `package-lock.json` file changes (indicating a change in dependencies).

- **Install dependencies**: The third step (`run: npm ci`) runs the `npm ci` command to install the project's dependencies from the `package-lock.json` file. This ensures a consistent dependency state across different environments.

- **Build the application**: The fourth step (`run: npm run build --if-present`) conditionally runs the `npm run build` command if it exists in the project's `package.json` file. This allows for flexibility in different project setups, where not all projects might have a `build` script defined.

However, it is important to note that if you clone the project from the course repository, you must add the following just after `runs-on: ubuntu-latest`:

```
defaults:
    run:
        working-directory: "./Chapter 9/getting-started-angular-tdd"
```

This final rendering is obtained by cloning the project from the book repository:

```
# This workflow will do a clean installation of node dependencies,
cache/restore them, build the source code and run tests across
different versions of node
# For more information see: https://docs.github.com/en/actions/
automating-builds-and-tests/building-and-testing-nodejs


name: Angular TDD CI/CD

on:
  push:
    branches: [ "main" ]
  pull_request:
    branches: [ "main" ]

jobs:
  build:

    runs-on: ubuntu-latest
    defaults:
      run:
        working-directory: "./Chapter 9/getting-started-angular-tdd"

    strategy:
      matrix:
        node-version: [18.x]
        # See supported Node.js release schedule at https://nodejs.
org/en/about/releases/

    steps:
    - uses: actions/checkout@v3
    - name: Use Node.js ${{ matrix.node-version }}
      uses: actions/setup-node@v3
      with:
        node-version: ${{ matrix.node-version }}
        cache: 'npm'
    cache-dependency-path: '**/package-lock.json'
      - run: npm ci
      - run: npm run build --if-present
```

Finally, if the process goes well, here's what you'll get in the GitHub Actions interface:

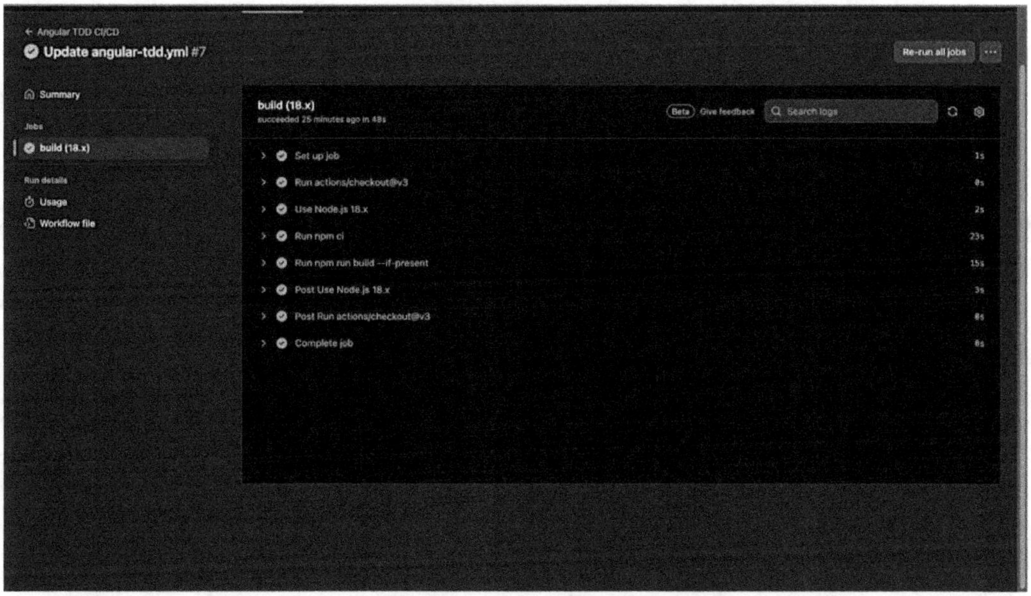

Figure 9.9 – Pipeline successfully completed

In the next section, we will learn how to set up CI/CD pipelines for automating test with GitHub Actions.

Setting up CI/CD pipelines for automating tests with GitHub Actions

In this new section, we'll update our previous workflow for running tests. Logic dictates that tests should be run before the build. Here's the test workflow:

```
test:
    runs-on: ubuntu-latest
    defaults:
      run:
        working-directory: './Chapter 9/getting-started-angular-tdd/'

    strategy:
      matrix:
        node-version: [18.x]
        # See supported Node.js release schedule at https://nodejs.
org/en/about/releases/
```

```yaml
    steps:
    - uses: actions/checkout@v3
    - name: Use Node.js ${{ matrix.node-version }}
      uses: actions/setup-node@v3
      with:
        node-version: ${{ matrix.node-version }}
        cache: 'npm'
        cache-dependency-path: '**/package-lock.json'
    - run: npm ci
    - run: npm run test
```

Now, let's combine the two workflows, test and build; here's how it looks in basic terms:

```yaml
name: Angular TDD CI/CD
on:
push:
    branches: [ "main" ]
pull_request:
    branches: [ "main" ]

jobs:
test:
    runs-on: ubuntu-latest
    defaults:
      run:
        working-directory: './Chapter 9/getting-started-angular-tdd/'

    strategy:
      matrix:
        node-version: [18.x]
        # See supported Node.js release schedule at https://nodejs.
org/en/about/releases/

    steps:
    - uses: actions/checkout@v3
    - name: Use Node.js ${{ matrix.node-version }}
      uses: actions/setup-node@v3
      with:
        node-version: ${{ matrix.node-version }}
        cache: 'npm'
        cache-dependency-path: '**/package-lock.json'
    - run: npm ci
    - run: npm run test
```

```
build:
    needs: test
    runs-on: ubuntu-latest
    defaults:
      run:
        working-directory: './Chapter 9/getting-started-angular-tdd/'

    strategy:
      matrix:
        node-version: [18.x]
        # See supported Node.js release schedule at https://nodejs.
org/en/about/releases/

    steps:
    - uses: actions/checkout@v3
    - name: Use Node.js ${{ matrix.node-version }}
      uses: actions/setup-node@v3
      with:
        node-version: ${{ matrix.node-version }}
        cache: 'npm'
        cache-dependency-path: '**/package-lock.json'
    - run: npm ci
    - run: npm run build --if-present
```

As mentioned in the preceding code block, the test workflow comes well before the build workflow. However, there is one aspect that immediately stands out. It's the repetition of many of the sequences found in the test and build. Based on the principle of **don't repeat yourself** (**DRY**), we're going to merge all the sequences into a single job, which we'll call test-and-build, for example. Here's how it looks:

```
# This workflow will do a clean installation of node dependencies,
cache/restore them, build the source code and run tests across
different versions of node
# For more information see: https://docs.github.com/en/actions/
automating-builds-and-tests/building-and-testing-nodejs

name: Angular TDD CI/CD

on:
  push:
    branches: [ "main" ]
  pull_request:
    branches: [ "main" ]
```

```
jobs:
  test-and-build:

    runs-on: ubuntu-latest
    defaults:
      run:
        working-directory: './Chapter 9/getting-started-angular-tdd/'

    strategy:
      matrix:
        node-version: [18.x]
        # See supported Node.js release schedule at https://nodejs.
org/en/about/releases/

    steps:
    - uses: actions/checkout@v3
    - name: Use Node.js ${{ matrix.node-version }}
      uses: actions/setup-node@v3
      with:
        node-version: ${{ matrix.node-version }}
        cache: 'npm'
        cache-dependency-path: '**/package-lock.json'
    - run: npm ci
    - run: npm run test --if-present
    - run: npm run build --if-present
```

Now, when we run the pipeline, we notice that the `npm run test --if-present` task runs in a block like this:

Figure 9.10 – Pipeline running

In fact, it's quite normal for us to have this problem because the `npm run test` executes the `ng test`. Since we're in an Angular project, it tries to launch Chrome in the pipeline. Unfortunately, it can't find it because we don't have a **graphical user interface (GUI)**. Hence, we get the following error:

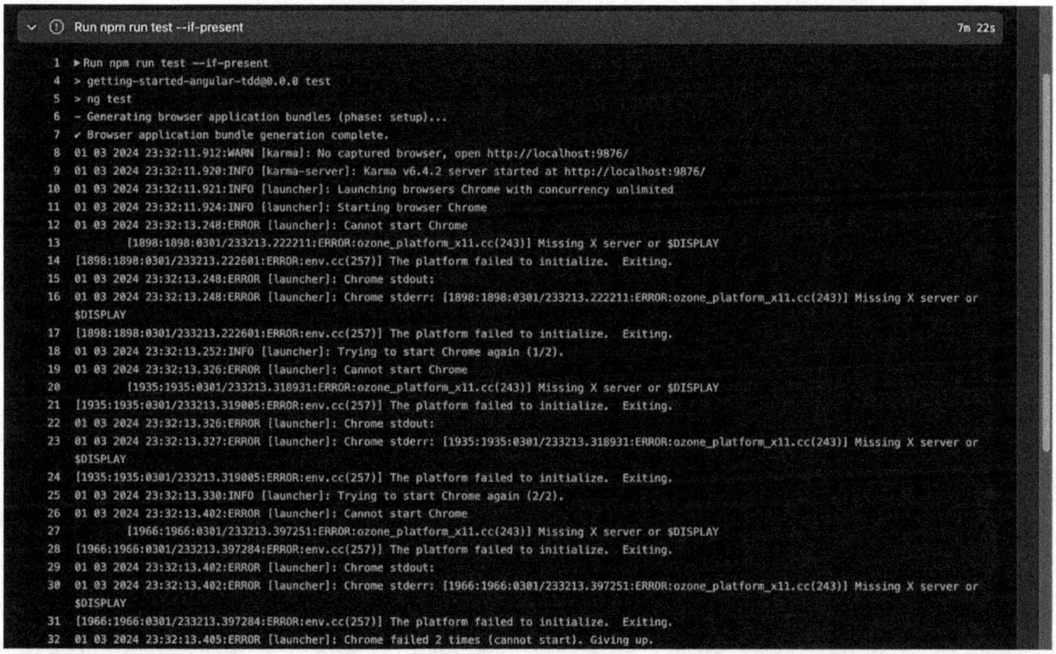

```
  ✓ ①  Run npm run test --if-present                                                                    7m 22s

   1   ▶ Run npm run test --if-present
   4   > getting-started-angular-tdd@0.0.0 test
   5   > ng test
   6   - Generating browser application bundles (phase: setup)...
   7   ✓ Browser application bundle generation complete.
   8   01 03 2024 23:32:11.912:WARN [karma]: No captured browser, open http://localhost:9876/
   9   01 03 2024 23:32:11.920:INFO [karma-server]: Karma v6.4.2 server started at http://localhost:9876/
  10   01 03 2024 23:32:11.921:INFO [launcher]: Launching browsers Chrome with concurrency unlimited
  11   01 03 2024 23:32:11.924:INFO [launcher]: Starting browser Chrome
  12   01 03 2024 23:32:13.248:ERROR [launcher]: Cannot start Chrome
  13        [1898:1898:0301/233213.222211:ERROR:ozone_platform_x11.cc(243)] Missing X server or $DISPLAY
  14   [1898:1898:0301/233213.222601:ERROR:env.cc(257)] The platform failed to initialize.  Exiting.
  15   01 03 2024 23:32:13.248:ERROR [launcher]: Chrome stdout:
  16   01 03 2024 23:32:13.248:ERROR [launcher]: Chrome stderr: [1898:1898:0301/233213.222211:ERROR:ozone_platform_x11.cc(243)] Missing X server or
       $DISPLAY
  17   [1898:1898:0301/233213.222601:ERROR:env.cc(257)] The platform failed to initialize.  Exiting.
  18   01 03 2024 23:32:13.252:INFO [launcher]: Trying to start Chrome again (1/2).
  19   01 03 2024 23:32:13.326:ERROR [launcher]: Cannot start Chrome
  20        [1935:1935:0301/233213.318931:ERROR:ozone_platform_x11.cc(243)] Missing X server or $DISPLAY
  21   [1935:1935:0301/233213.319005:ERROR:env.cc(257)] The platform failed to initialize.  Exiting.
  22   01 03 2024 23:32:13.326:ERROR [launcher]: Chrome stdout:
  23   01 03 2024 23:32:13.327:ERROR [launcher]: Chrome stderr: [1935:1935:0301/233213.318931:ERROR:ozone_platform_x11.cc(243)] Missing X server or
       $DISPLAY
  24   [1935:1935:0301/233213.319005:ERROR:env.cc(257)] The platform failed to initialize.  Exiting.
  25   01 03 2024 23:32:13.330:INFO [launcher]: Trying to start Chrome again (2/2).
  26   01 03 2024 23:32:13.402:ERROR [launcher]: Cannot start Chrome
  27        [1966:1966:0301/233213.397251:ERROR:ozone_platform_x11.cc(243)] Missing X server or $DISPLAY
  28   [1966:1966:0301/233213.397284:ERROR:env.cc(257)] The platform failed to initialize.  Exiting.
  29   01 03 2024 23:32:13.402:ERROR [launcher]: Chrome stdout:
  30   01 03 2024 23:32:13.402:ERROR [launcher]: Chrome stderr: [1966:1966:0301/233213.397251:ERROR:ozone_platform_x11.cc(243)] Missing X server or
       $DISPLAY
  31   [1966:1966:0301/233213.397284:ERROR:env.cc(257)] The platform failed to initialize.  Exiting.
  32   01 03 2024 23:32:13.405:ERROR [launcher]: Chrome failed 2 times (cannot start). Giving up.
```

Figure 9.11 – Pipeline fails

To fix this, we're going to make a few changes to the project, in particular to the `angular.json` file, by adding a configuration to the test configuration like this:

```
"configurations": {
        "ci": {
            "watch": false,
            "progress": false,
            "browsers": "ChromeHeadlessCI"
    }
}
```

This is the complete test configuration:

```
"test": {
        "builder": "@angular-devkit/build-angular:karma",
        "options": {
          "polyfills": [
```

```
              "zone.js",
              "zone.js/testing"
          ],
          "tsConfig": "tsconfig.spec.json",
          "inlineStyleLanguage": "scss",
          "assets": [
            "src/favicon.ico",
            "src/assets"
          ],
          "styles": [
            "src/styles.scss"
          ],
          "scripts": []
        },
        "configurations": {
        "ci": {
          "watch": false,
          "progress": false,
          "browsers": "ChromeHeadlessCI"
        }
      }
    }
```

After that, we need to create a `karma.conf.js` file in the `src` folder of our Angular project, if the file doesn't already exist. In this file, we'll put the source code related to Karma's configuration:

```
// Karma configuration file, see link for more information
// https://karma-runner.github.io/1.0/config/configuration-file.html

process.env.CHROME_BIN = require("puppeteer").executablePath();

module.exports = function (config) {
  config.set({
    basePath: "",
    frameworks: ["jasmine", "@angular-devkit/build-angular"],
    plugins: [
      require("karma-jasmine"),
      require("karma-chrome-launcher"),
      require("karma-jasmine-html-reporter"),
      require("karma-coverage-istanbul-reporter"),
      require("@angular-devkit/build-angular/plugins/karma"),
    ],
    client: {
      clearContext: false, // leave Jasmine Spec Runner output visible
in browser
```

```
    },
    coverageIstanbulReporter: {
      dir: require("path").join(__dirname, "../coverage"),
      reports: ["html", "lcovonly"],
      fixWebpackSourcePaths: true,
    },
    reporters: ["progress", "kjhtml"],
    port: 9876,
    colors: true,
    logLevel: config.LOG_INFO,
    autoWatch: true,
    browsers: ["Chrome"],
    customLaunchers: {
      ChromeHeadlessCI: {
        base: "ChromeHeadless",
        flags: ["--no-sandbox", "--disable-gpu"],
      },
    },
    singleRun: false,
  });
};
```

Next, we need to install two packages in dev mode, namely `puppeteer` and `karma-coverage-istanbul-reporter`, by doing the following:

```
$ npm i --save-dev puppeteer karma-coverage-istanbul-reporter
```

Finally, in our GitHub Actions pipeline, we replace npm `run test -if-present` with npm `run test -- --configuration=ci`, and here's the result:

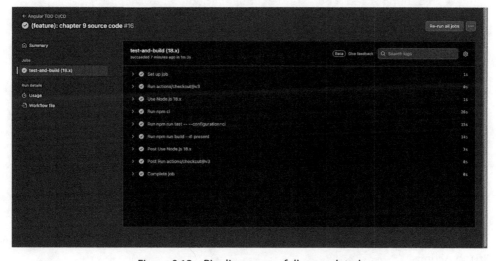

Figure 9.12 – Pipeline successfully completed

Well done! In the following, you'll find a summary of the workflow with all the changes made to date, along with the associated source code:

```
# This workflow will do a clean installation of node dependencies,
cache/restore them, build the source code and run tests across
different versions of node
# For more information see: https://docs.github.com/en/actions/
automating-builds-and-tests/building-and-testing-nodejs


name: Angular TDD CI/CD

on:
  push:
    branches: [ "main" ]
  pull_request:
    branches: [ "main" ]

jobs:
  test-and-build:

    runs-on: ubuntu-latest
    defaults:
      run:
        working-directory: './Chapter 9/getting-started-angular-tdd/'

    strategy:
      matrix:
        node-version: [18.x]
        # See supported Node.js release schedule at https://nodejs.
org/en/about/releases/

    steps:
    - uses: actions/checkout@v3
    - name: Use Node.js ${{ matrix.node-version }}
      uses: actions/setup-node@v3
      with:
        node-version: ${{ matrix.node-version }}
        cache: 'npm'
        cache-dependency-path: '**/package-lock.json'
    - run: npm ci
    - run: npm run test -- --configuration=ci
    - run: npm run build --if-present
```

In the next section, we will learn how to set up CI/CD pipelines for automating deployment processes with GitHub Actions.

Setting up CI/CD pipelines for automating the deployment process with GitHub Actions

CD is the practice of automatically deploying changes to production as soon as they have passed through the production pipeline. This includes the automated processes of testing, building, and deploying. GitHub Actions supports CD, enabling you to automate these processes efficiently.

We're not going to do it on this project, but we'll see how we can. To begin with, this will be a test and build stage like the other two. Deployment takes place naturally at the end of development, so the same will apply to the workflow. At the end of our current workflow, just after the build, we'll add the following to deploy on a remote server:

```
- name: Upload build files to remote server
  uses: appleboy/scp-action@master
  with:
    host: ${{ secrets.SI_HOST }}
    username: ${{ secrets.SI_USERNAME }}
    password: ${{ secrets.SI_PASSWORD }}
    port: ${{ secrets.SI_PORT }}
    source: "[SORUCE_FOLDER]"
    target: "[DESTINATION_TARGET_ON_YOUR_SERVER]"
```

The provided GitHub Actions workflow snippet is designed to automate the process of uploading build files to a remote server. This is a common step in a CI/CD pipeline for deploying applications. Here's a breakdown of the workflow step:

- `appleboy/scp-action@master`: This specifies that this step uses the `scp-action` action from the `appleboy` GitHub repository. This action is designed to securely copy files from your GitHub Actions runner to a remote server using the **secure copy protocol** (**SCP**). The `@master` tag indicates that the action should use the code from the master branch of the repository.

- `host`: The address of the remote server where the files will be uploaded. This value is retrieved from a GitHub secret named `SI_HOST`.

- `username`: The username for authenticating with the remote server. This value is retrieved from a GitHub secret named `SI_USERNAME`.

- `password`: The password for authenticating with the remote server. This value is retrieved from a GitHub secret named `SI_PASSWORD`.

- `port`: The port number for connecting to the remote server. This value is retrieved from a GitHub secret named `SI_PORT`.

- `source`: The path to the files that will be uploaded. In this case, it's set to upload all files in the source build directory.

- `target`: The destination path on the remote server where the files will be uploaded.

For other cloud-oriented platforms, GitHub Actions makes the task easier by offering deployment templates for most of these platforms.

Summary

In summary, this chapter covered the fundamental concepts of CI and CD practices within the SDLC, highlighting their importance and benefits. We began by understanding that CI is a process that aims to automate the integration of code changes into a shared repository, facilitating automated builds and tests to quickly detect and correct problems. This practice is crucial for the early detection of bugs and errors, promoting a faster feedback loop, encouraging collaboration, and improving code quality.

CD is then introduced as an extension of CI, focusing on automating the deployment process while ensuring that the software is always in a releasable state. This practice enables rapid and reliable releases of software in different environments, reducing the risk of deployment errors and enabling faster time-to-market.

This chapter also looked at the practical aspects of setting up CI/CD pipelines using GitHub Actions, a popular workflow automation tool. It explained how to automate the build process, including installing dependencies, compiling code, and running tests, as well as automating the deployment process.

Key concepts and practices of continuous processes were examined, including the importance of making small, iterative changes, adopting trunk-based development, maintaining rapid build and test phases, and decoupling deployment from production release. These practices are essential for establishing efficient, reliable continuous processes that accelerate development cycles, improve software quality, and deliver value to customers more quickly.

In addition, the chapter discussed the role of testing in CI/CD processes, highlighting the importance of different types of testing, such as smoke tests, unit tests, integration tests, system tests, and acceptance tests. These tests are essential to guarantee software quality and stability, providing rapid feedback on the state of the code base and helping to detect and correct problems early in the development process.

In the next chapter, we'll learn about the best practices and patterns for test-driven development.

Best Practices and Patterns for Angular TDD

Test-driven development (**TDD**) in Angular is a methodology that emphasizes writing tests before the actual implementation code. This approach ensures that code is thoroughly tested and meets specified requirements. In the fast-changing landscape of frontend development, where frameworks such as Angular enable complex enterprise-level applications to be built, the importance of good testing cannot be overstated. TDD results in high-quality production code and a robust code base, ensuring that the application works as expected and can stand the test of time.

Angular developers often find themselves navigating the intricacies of unit testing, particularly when dealing with components, services, and pipes. The process of writing unit tests for Angular applications involves understanding the structure of Angular unit tests, the need for a dummy module for testing, and the iterative process of building an Angular service using TDD. This iterative process, in which developers alternate between writing tests and the smallest amount of code necessary to pass the test, is the cornerstone of TDD. It allows functionality to be built in small, verifiable increments, ensuring that each piece of functionality is thoroughly tested.

To implement TDD effectively in Angular, developers need to be aware of common pitfalls, such as writing tests that are too dependent on implementation details. This can lead to brittle tests that break with each refactoring, even if the external behavior remains consistent. By focusing on the desired behavior rather than the specific implementation, developers can write clear, precise expectations that guide the development process. In addition, the iterative process of writing tests and code implementation helps to build functionality in small, verifiable increments, leading to a comprehensive test suite that articulates the responsibilities and expected behaviors of the system's parts.

The following topics will be covered in this chapter:

- Best practices for TDD in Angular projects
- Exploring patterns for implementing TDD in any Angular project
- Choosing the right TDD pattern for your Angular project

Best practices for TDD in Angular projects

TDD is a methodology that prioritizes test creation before implementation coding. This strategy ensures that code is thoroughly tested and aligned with defined requirements. Here are some recommended practices and models for implementing TDD in Angular:

- **Writing tests before implementation**: Start by writing a test for a feature or functionality that doesn't exist yet. This test will initially fail because the feature hasn't been implemented yet. The test should focus on the desired behavior rather than the specific implementation, ensuring that the test is clear and concise.

- **Use TestBed to test Angular services**: Angular services are injectable classes that contain business logic. Testing these services is crucial to ensuring the functionality of your application. Use `TestBed.configureTestingModule` to create a mock module for testing, and `TestBed.inject` to initialize the service in this module. This configuration ensures that the service is isolated from external dependencies, enabling accurate testing.

- **Avoid implementation details in tests**: Tests should verify the behavior of a feature from the perspective of an end user or an API consumer, without assuming knowledge about the internal workings of the feature. This approach helps in creating tests that are resilient to changes in the implementation.

- **Iterate between writing tests and implementation**: Toggle between writing tests and the smallest amount of code necessary to pass the test. This iterative process helps in building functionality in small, verifiable increments, ensuring that each piece of functionality is tested thoroughly.

- **Use mocks, spies, and stubs**: To ensure that tests are not coupled with external dependencies, use mocks, spies, and stubs. Mocks provide controlled replacement implementations for dependencies, spies track function calls, and stubs offer predetermined behavior. These tools help in isolating the component or service being tested from external factors.

- **Test component interactions and state management**: For components, test both internal state changes and interactions with global state stores. Use `RouterTestingModule` to simulate navigation events and verify that components react to route changes as expected. For interactions between components, simulate scenarios where components communicate via inputs and outputs using `TestBed`.

- **Maintain and refactor tests**: Regularly review and refactor tests to ensure they remain relevant and reflective of the current state of the application. Refactor tests in tandem with the application code, ensuring that tests undergo the same rigor of improvement as your production code. Use version control strategies that include test updates as part of the feature branch to catch breaking changes early.

- **Optimize test performance**: Optimize the performance of your unit tests by grouping tests logically, using debounce and throttle techniques where applicable, and efficiently handling dependencies. Utilize Angular's hierarchical injector to provide service mocks at the right level, reducing redundancy across tests. Regularly audit the test suite to remove obsolete tests and refactor those that can be merged or simplified.

- **Architect resilient Angular unit tests**: Design your tests to be independent of implementation details, focusing on the component or service's public API. Use `beforeEach()` blocks effectively to set up the necessary conditions for each test without side effects on other tests. Write tests that allow components or services to expand according to the application's requirements without necessitating constant test rewrites.

- **Continuous improvement**: Continuously refine both production and test code bases, ensuring that your tests are as maintainable and efficient as the features they validate. Reflect on how your tests might need to adapt to your business logic correctly to represent the evolving application, ensuring your tests remain robust and representative.

In summary, with these best practices, Angular developers can foster a sustainable test culture that adapts to change while keeping quality at the forefront. In the next section, we will explore patterns for implementing TDD in any Angular project.

Exploring patterns for implementing TDD in any Angular project

Exploring patterns for implementing TDD in Angular projects is a critical step toward ensuring the robustness and reliability of your applications. TDD is a methodology that involves writing tests before the actual implementation code, ensuring that the code is thoroughly tested and meets the specified requirements. This approach not only leads to high-quality production code but also fosters a culture of continuous improvement and resilience against changes.

TDD patterns are strategies or approaches to implementing TDD within the context of Angular projects. These patterns can vary based on the specific needs of the project, the complexity of the application, and the team's familiarity with TDD practices. Some of the most common TDD patterns are as follows:

- **Unit testing**: This pattern focuses on testing individual components or services in isolation. It is essential as it ensures that each part of your application functions as expected in isolation from the rest of the system.

- **Component testing**: Angular applications are built around components, making them a fundamental part of the user interface. When implementing TDD in Angular projects, starting with testing components ensures that they render correctly, handle user interactions, and update the UI as expected.

- **Service testing**: Services in Angular encapsulate business logic and data manipulation functions. Writing tests for services ensures that they perform their intended functions, interact correctly with external resources, and handle errors gracefully.

- **Integration testing**: This pattern involves testing the interaction between different parts of the application, such as components and services. It helps in identifying issues that may arise when different parts of the application work together.

- **End-to-end (E2E) testing**: This pattern simulates user interactions with the application in a real-world scenario, testing the entire application flow from start to finish. It is crucial as it ensures that the application behaves as expected from the user's perspective.

In summary, exploring patterns for implementing TDD in Angular projects is a journey toward building high-quality, scalable, and maintainable applications. By adopting TDD best practices and patterns, developers can enhance the efficiency of their development process, improve code quality, and deliver exceptional user experiences. As the software development landscape continues to evolve, the integration of TDD in Angular projects serves as a cornerstone for fostering a culture of excellence and continuous improvement.

In the next section, we'll learn how to choose a TDD pattern for your Angular project.

Choosing a TDD pattern for your Angular project

Choosing a TDD pattern for your Angular project is crucial for ensuring the reliability, maintainability, and scalability of your application. As mentioned previously, TDD is a methodology that involves writing tests before the actual implementation code, ensuring that the code is thoroughly tested and meets the specified requirements. However, the choice of TDD pattern can significantly impact the development process, testing strategy, and overall project outcomes. This section will help you understand the TDD patterns that are available and why they might be appropriate for your Angular project. Choosing a TDD pattern for your Angular project depends on several factors:

- **Project complexity**: For complex applications with numerous components and services, a combination of unit, integration, and E2E testing patterns may be necessary. This approach ensures that each part of the application is thoroughly tested in isolation, as well as in the context of the entire application.

- **Team expertise**: The team's familiarity with TDD practices and the specific testing tools and frameworks available for Angular can influence the choice of TDD pattern. For example, Angular provides robust testing tools and libraries that facilitate both unit and integration testing.

- **Project requirements**: The specific requirements of your project, such as performance, security, and user experience, can also guide the choice of TDD pattern. For instance, E2E testing is particularly useful for projects that require a high level of user interaction and real-world testing.

In summary, choosing a TDD pattern for your Angular project is a strategic decision that should be based on the project's complexity, the team's expertise, and the specific requirements of the application.

Summary

In this chapter, we explored best practices and patterns for effectively implementing TDD in Angular projects. TDD encourages the process of writing tests before the actual code, guaranteeing clear functionality and reducing the risk of regression.

Key lessons included the TDD cycle with its red-green-refactor phases and models such as the **Arrange-Act-Assert** (**AAA**) structure for organizing tests and efficiently simulating dependencies. By following these patterns and best practices, such as writing focused tests, continuous refactoring, test automation, and fostering collaboration within the team, developers can enhance the quality of their code, streamline the development process, and deliver better user experiences. Embracing TDD in Angular projects is not just a technique but a mindset that fosters a culture of continuous improvement and excellence in software development.

In the next chapter, we will learn how to refactor and improve Angular code through TDD.

11

Refactoring and Improving Angular Code through TDD

Refactoring your Angular code using **test-driven development** (TDD) is a systematic and effective approach to improving the quality of your code. TDD involves writing tests before writing the actual code, thus ensuring that the code addresses the desired requirements and is robust, maintainable, and reliable. This methodology is particularly beneficial for Angular applications, where it ensures that the code is well structured, efficient, and easy to maintain.

In this section, we delve into the significance of adopting a "tests first" strategy, exploring the advantages of TDD during refactoring processes, selecting the optimal tests to create, understanding what constitutes **code smells**, emphasizing the importance of addressing code smells within Angular projects, and identifying prevalent code smells in Angular applications.

In summary, here are the main topics that will be covered in this chapter:

- Refactoring Angular code through TDD
- Identifying code smells and areas for improvement in Angular applications
- Iterative improvement – red-green-refactor cycle for continuous code enhancement

Technical requirements

To follow along with the examples and exercises in this chapter, you will need to have a basic understanding of Angular and TypeScript, as well as the following technical requirements:

- Node.js and npm installed on your computer
- Angular CLI installed globally
- A code editor, such as Visual Studio Code, installed on your computer

The code files required for this chapter can be found at `https://github.com/PacktPublishing/Mastering-Angular-Test-Driven-Development/tree/main/Chapter%2011`.

Refactoring Angular code with TDD

Refactoring existing code in complex Angular applications can be a nerve-wracking task. You want to improve the code's structure and organization without accidentally breaking existing functionalities. This is where TDD comes into play, offering a structured approach to navigate refactoring with confidence. In this section, we'll see the power of the test-first approach, the benefits of TDD in refactoring, and how to choose the right tests to write.

The power of the test-first approach

TDD flips the traditional coding script. Instead of writing code first and then testing it, TDD emphasizes writing tests before making any modifications to the code. These tests essentially define the expected behavior of the code you intend to refactor. Here's how it works:

- **The red state**: You start by writing a test for a specific functionality within the code you want to refactor. This initial test will likely fail, signifying that the desired behavior isn't yet implemented. This *red* state acts as a starting point.

- **The green state**: With the failing test as your guide, you write just enough code to make the test pass. This initial implementation might be basic, but the focus is on ensuring it accurately reflects the intended behavior defined by the test. Now, the test is in a *green* state, indicating successful implementation.

- **Refactor**: Here's where the magic happens. With the safety net of a passing test, you can now refactor the code to improve its readability, maintainability, and efficiency. This might involve the following:

  - Breaking down long methods into smaller, well-defined functions

  - Extracting reusable components or services from large components

  - Applying design patterns for better code organization

  - Simplifying logic to enhance clarity

Throughout this refactoring stage, the passing test ensures that these changes don't introduce any unintended side effects. Essentially, you're improving the code's internal workings without altering its external behavior as verified by the test. As an example, let's start with this block of code from our `calculator.component.ts` component in the Chapter 9 folder (https://github.com/PacktPublishing/Mastering-Angular-Test-Driven-Development/tree/main/Chapter%209/getting-started-angular-tdd):

```
calculate(): void {
    if (this.calculatorForm.get('operator')?.value === '+') {
      this.add(
        this.calculatorForm.get('operand1')?.value,
        this.calculatorForm.get('operand2')?.value
      );
    }

    if (this.calculatorForm.get('operator')?.value === '-') {
      this.substract(
        this.calculatorForm.get('operand1')?.value,
        this.calculatorForm.get('operand2')?.value
      );
    }

    if (this.calculatorForm.get('operator')?.value === '*') {
      this.multiply(
        this.calculatorForm.get('operand1')?.value,
        this.calculatorForm.get('operand2')?.value
      );
    }

    if (this.calculatorForm.get('operator')?.value === '/') {
      this.divide(
        this.calculatorForm.get('operand1')?.value,
        this.calculatorForm.get('operand2')?.value
      );
    }
}
```

Here is the source code for the tests corresponding to this function in our `calculator.component.spec.ts` file:

```
  it('should be valid when all of the fields are filled in correctly',
  () => {
    calculator.calculatorForm.get('operand1')?.setValue(123);
    calculator.calculatorForm.get('operand2')?.setValue(456);
```

```
    calculator.calculatorForm.get('operator')?.setValue('+');

    expect(calculator.calculatorForm.valid).toBe(true);
  });

  it('should be invalid when one of the field is not filled in
correctly', () => {
    calculator.calculatorForm.get('operand1')?.setValue(123);
    calculator.calculatorForm.get('operator')?.setValue('+');

    expect(calculator.calculatorForm.valid).toBe(false);
  });

  it('should add when the + operator is selected and the calculate
button is clicked', () => {
    calculator.calculatorForm.get('operand1')?.setValue(2);
    calculator.calculatorForm.get('operand2')?.setValue(3);
    calculator.calculatorForm.get('operator')?.setValue('+');
    calculator.calculate();
    expect(calculator.result).toBe(5);
  });

  it('should subtract when the - operator is selected and the
calculate button is clicked', () => {
    calculator.calculatorForm.get('operand1')?.setValue(2);
    calculator.calculatorForm.get('operand2')?.setValue(3);
    calculator.calculatorForm.get('operator')?.setValue('-');
    calculator.calculate();
    expect(calculator.result).toBe(-1);
  });

  it('should multiply when the * operator is selected and the
calculate button is clicked', () => {
    calculator.calculatorForm.get('operand1')?.setValue(2);
    calculator.calculatorForm.get('operand2')?.setValue(3);
    calculator.calculatorForm.get('operator')?.setValue('*');
    calculator.calculate();
    expect(calculator.result).toBe(6);
  });

  it('should divide when the / operator is selected and the
calculation button is clicked.', () => {
    calculator.calculatorForm.get('operand1')?.setValue(3);
    calculator.calculatorForm.get('operand2')?.setValue(2);
    calculator.calculatorForm.get('operator')?.setValue('/');
```

```
    calculator.calculate();
    expect(calculator.result).toBe(1.5);
  });
```

As a reminder, all tests are green, as shown in the following screenshot of Karma launched on a browser:

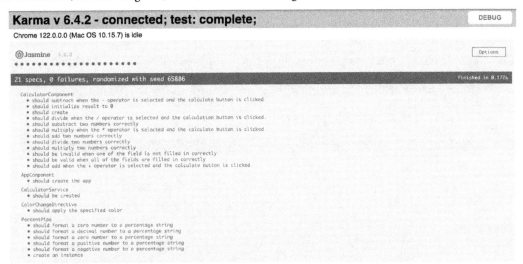

Figure 11.1 – Calculator component test succeeded in the terminal

We will now refactor the code of our `calculate()` function in the `calculator.component.ts` component as follows:

```
calculate(): void {
  const operator = this.calculatorForm.get('operator')?.value;
  const operand1 = this.calculatorForm.get('operand1')?.value;
  const operand2 = this.calculatorForm.get('operand2')?.value;

  if (!operator ||!operand1 ||!operand2) return;
  switch (operator) {
    case '+':
      this.add(operand1, operand2);
      break;
    case '-':
      this.subtract(operand1, operand2);
      break;
    case '*':
      this.multiply(operand1, operand2);
      break;
    case '/':
```

```
        this.divide(operand1, operand2);
        break;
    default:
        console.error(`Unsupported operator: ${operator}`);
        break;
    }
}
```

Now, what's going on with our tests? They're still all green as the following screenshot shows:

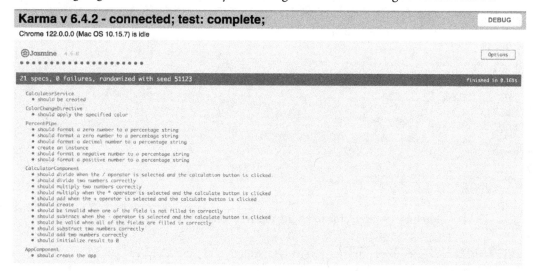

Figure 11.2 – Green tests

This means that our refactoring is correct because if it wasn't, our tests would be red. We now understand the importance of TDD, because we can concentrate on refactoring our methods with peace of mind. In the next section, we'll learn some benefits of TDD in refactoring.

Benefits of TDD in refactoring

This section delves into the rewarding benefits of utilizing TDD during Angular code refactoring. Here are some of them:

- **Increased confidence**: The passing test acts as a safety net, allowing you to experiment with different refactoring techniques without worrying about breaking existing functionality. This boosts your confidence during the process.

- **Improved design**: Thinking about the tests first encourages you to write modular code with well-defined functions. This leads to cleaner and more maintainable code in the long run.

- **Enhanced maintainability**: A comprehensive test suite becomes a living documentation of the code's expected behavior. This simplifies future modifications and bug fixes, as you can rely on the tests to catch regressions.

- **Better code coverage**: TDD naturally encourages you to focus on covering various code paths with tests. This results in more robust applications with fewer hidden bugs.

Examples in action

Let's see how TDD can be applied to common refactoring scenarios in Angular:

- **Refactoring a long service method**: Imagine a service method responsible for fetching and processing a massive amount of data. You can write a test that focuses on a specific aspect of the data processing logic. Initially, the test would fail. Then, you'd refactor the service method to extract that logic into a separate, well-tested function. This improves code readability and maintainability while the test ensures the core functionality remains intact.

- **Transforming a god component**: A "god component" refers to an overly complex component, often violating the single responsibility principle by handling too many responsibilities or functionalities within itself. This term is used to highlight the negative aspects of having components that grow too large and become difficult to manage, test, and understand. Such components tend to inject numerous services, perform multiple tasks, and can lead to significant maintenance challenges over time. This type of component performing a large number of tasks can therefore be redesigned by creating services dedicated to specific functionalities. Tests can be written to verify the behavior of the refactored component and the newly created services.

Choosing the right tests to write

When prioritizing tests for refactoring with TDD, consider these strategies:

- **Focus on pain points**: Start with functionalities that are causing problems in your codebase, such as areas prone to errors or difficult to understand.

- **Start small**: Begin with smaller, well-defined tests that target specific functionalities. This allows for quicker iterations through the red-green-refactor cycle.

- **Test integration points**: When refactoring components that interact with services or other components, write tests that verify these interactions alongside the component itself.

In the next section, we'll explain how to identify code smells and areas for improvement in Angular applications.

Identifying code smells and areas for improvement in Angular applications

While your Angular application might seem functional on the surface, there could be underlying issues waiting to erupt. These issues, known as code smells, don't necessarily cause immediate problems but indicate areas in your codebase that could benefit from refactoring. Just like a cluttered room can be stressful and unproductive, smelly code can make it difficult to maintain, understand, and extend your application. This section delves into the world of code smells in Angular applications. We'll explore what they are, why they matter, and how to identify them proactively. By understanding these code smells, you'll be equipped to prioritize refactoring efforts using TDD, ultimately leading to a cleaner, more maintainable, and robust code base.

What are code smells?

Imagine walking into a kitchen where dirty dishes are piled up high in the sink, spices are scattered across the counter, and expired food lingers in the fridge. This unpleasant scene might not prevent you from cooking a basic meal, but it certainly wouldn't be an enjoyable or efficient experience. Code smells are analogous to this messy kitchen in the software development world.

Coined by Martin Fowler in his book, *Refactoring: Improving the Design of Existing Code*, code smells are indicators of potential problems within your code base. They don't necessarily represent functional bugs that cause the application to crash. Instead, they signify areas that could be improved for better readability, maintainability, and long-term health of your code.

Code smells are not bugs, but they can attract bugs in the future. They act like red flags, warning you of potential trouble spots that could become problematic as your application evolves.

Why should we care about code smells in Angular?

Ignoring a messy kitchen might lead to unpleasant odors, fruit flies, and frustration when you need to cook a meal. Similarly, neglecting code smells in your Angular application can have several negative consequences:

- **Decreased maintainability**: Smelly code becomes difficult to understand and modify over time. As your application grows and features are added, the complexity of tangled code can make changes cumbersome and error prone.

- **Increased debugging time**: When bugs arise in smelly code, it can be challenging to pinpoint the root cause. The lack of clear structure and organization makes it like searching for a needle in a haystack, wasting valuable developer time.

- **Reduced team productivity**: Working with smelly code can be frustrating and demotivating for developers. The cognitive overhead of deciphering tangled logic slows down development and hinders collaboration.

- **Technical debt**: Unattended code smells accumulate over time, creating a technical debt that needs to be addressed eventually. This debt can become a significant burden, requiring dedicated resources and potentially delaying new feature development.

By proactively identifying and refactoring code smells, you can do the following:

- **Improve code readability**: Clean and well-structured code is easier to understand for both you and other developers working on the project. This reduces onboarding time for new team members and fosters better collaboration.

- **Enhance maintainability**: Refactored code is easier to modify and adapt as your application's requirements evolve. This allows you to introduce new features and bug fixes more efficiently.

- **Reduce debugging time**: Cleaner code with a clear separation of concerns makes it easier to isolate and fix problems when bugs arise.

- **Boost team productivity**: Working with well-structured code improves developer experience and satisfaction. This leads to higher productivity and a more positive development environment.

- **Minimize technical debt**: By addressing code smells early on, you prevent them from accumulating and becoming a significant burden in the future.

In essence, prioritizing code smell refactoring is an investment in the long-term health and maintainability of your Angular application.

Identifying the most common code smells in Angular applications

Now that we understand the importance of identifying code smells, let's take a look at some of the most common offenses you might encounter in your Angular application:

- **The long and winding method**: Imagine a method in your service that stretches across dozens of lines, handling various tasks. This is a prime example of a long method, a code smell suggesting a lack of modularity. These methods can be difficult to understand, test, and modify. Refactoring involves breaking down such behemoths into smaller, well-defined functions, each focusing on a specific task. This enhances code readability and maintainability.

- **The god component**: Have you encountered a component overloaded with responsibilities? This is a "god component," handling everything from data fetching to complex UI logic. Such components become maintenance nightmares as changes in one area can ripple through the entire component, causing unintended consequences. Refactoring can involve the following:

- **Creating dedicated services**: Extract functionalities related to data access, business logic, or calculations into separate services. These services can be reused by multiple components, promoting better organization.

  - **Splitting the component**: Break down the god component into smaller, more focused components, each handling a specific aspect of the UI or functionality.

- **The code duplication monster**: Seeing the same block of code copy-pasted across different parts of your application? This code duplication not only wastes space but also makes maintenance a challenge. Any bug fix in one instance needs to be replicated across all copies. Refactoring involves identifying these repetitive code snippets and creating the following:

  - **Reusable components**: If the duplicated code deals with UI elements, consider creating a reusable component that can be used in multiple places.

  - **Services or utility functions**: For duplicated logic unrelated to the UI, extract them into services or utility functions that can be shared across components.

  - **The magic number mystery**: Scattered numeric constants with no clear meaning throughout your code base are like magic tricks; they are confusing and difficult to maintain. Imagine having a constant 10 used for pagination but its purpose is unclear. Refactoring involves replacing these magic numbers with named variables or constants. For example, use ITEMS_PER_PAGE instead of 10, making the code more self-documenting and easier to understand.

- **The spaghetti code maze**: Imagine code that winds and twists, lacking clear structure and organization. This is spaghetti code, making it a challenge to navigate, understand, and modify. TDD can be a powerful tool to combat spaghetti code. By writing tests first and then refactoring the code to meet those tests, you can introduce structure and improve the overall organization of your code base.

In the next section, we will learn about iterative improvement: the red-green-refactor cycle for continuous code enhancement.

Iterative improvement – red-green-refactor cycle for continuous code enhancement

Refactoring existing code in Angular applications can be a daunting task. You want to improve the code's structure and organization, but the fear of introducing regressions (bugs) often looms large. This is where TDD steps in, offering a structured and iterative approach to navigate refactoring with confidence. At the heart of TDD lies the "red-green-refactor" cycle, a powerful technique for making incremental improvements to your code base while ensuring its functionality remains intact.

Imagine you're a sculptor working on a large block of marble. The red-green-refactor cycle is like your roadmap to transforming that raw material into a masterpiece. Here's a breakdown of each stage and its significance in the context of refactoring Angular code:

Red – setting the stage with failing tests

The cycle begins with red, signifying a failing test. This might seem counterintuitive – why write a test that's destined to fail? The purpose of this initial red test is to define the desired behavior of the code you intend to refactor. Think of it as a blueprint outlining the functionality you want to achieve. Here's what creating a red test entails:

- **Identify the refactoring target**: Start by pinpointing a specific area of your Angular application that exhibits code smells or requires improvement. This could be a long method in service, a god component handling numerous tasks, or duplicated code snippets.

- **Define the expected outcome**: Clearly outline what the refactored code should do. What data should it process? How should it interact with the UI? Write a test that reflects this expected behavior. Remember, this test will initially fail, as the desired functionality isn't implemented yet.

A failing red test serves a critical purpose. It establishes a baseline – a clear understanding of the functionality that's currently missing. This provides a safety net during the refactoring process. As you make code changes, the failing test ensures you're on the right track and haven't accidentally broken existing functionalities.

Green – making the test pass with minimal code

Once you have your red test in place, it's time to move to the green stage. Here, the objective is to write just enough code to make the failing test pass. Don't get caught up in writing perfectly optimized or elegant code at this point. Focus on the core functionality defined by the test. Here are some key considerations for the green stage:

- **Simple implementations**: The initial code you write to make the test pass might be basic. It doesn't have to be the most efficient or well-structured solution yet. The priority is to get the test passing and establish a baseline for refactoring.

- **Focus on functionality**: Ensure the code you write fulfills the specific behavior outlined in the test. Don't introduce unnecessary features or logic at this stage.

A passing green test signifies a crucial milestone. It verifies that the core functionality you're refactoring is now implemented, albeit potentially in a basic form. This green test acts as a safety net throughout the refactoring process. As you make further code changes, you can rely on the test to ensure you haven't strayed from the desired outcome.

Refactor – transforming the code with confidence

With a passing green test as your safety net, you've reached the heart of the cycle – the refactor stage. Here's where you can unleash your refactoring skills to improve the code's readability, maintainability, and efficiency. Here are some potential areas you might want to focus on during refactoring:

- **Modularize long methods**: Break down those long, monolithic methods into smaller, well-defined functions. This enhances code readability and makes it easier to understand the logic flow.

- **Extract reusable components and services**: If your component has become a god component handling numerous tasks, consider extracting functionalities into dedicated services or reusable components. This promotes better organization and separation of concerns.

- **Eliminate duplication**: Identify and refactor repetitive code snippets into reusable components, services, or utility functions. This reduces code redundancy and simplifies maintenance.

- **Apply design patterns**: Consider incorporating design patterns that promote better code structure and organization. This can make your code more maintainable and easier to understand for other developers.

- **Simplify logic**: Look for opportunities to streamline complex logic and enhance code clarity. This can involve using more descriptive variable names, breaking down complex conditional statements, or utilizing helper functions.

Throughout the refactoring process, keep the green test in mind. Repeat this cycle iteratively, tackling one aspect of the code at a time. Each completed cycle leaves you with cleaner, more maintainable code, with robust tests guaranteeing its continued functionality. The emphasis on small, incremental changes promotes a more controlled, less error-prone refactoring process.

Summary

In conclusion, refactoring and improving Angular code through TDD is a powerful approach to enhancing the quality, maintainability, and efficiency of your Angular applications. By following the TDD methodology, developers can ensure that their code is robust, well structured, and easy to understand. This approach not only helps in identifying and addressing code smells but also in developing a solid foundation for future enhancements and modifications. TDD encourages developers to write tests before writing the actual code, ensuring that the code meets the defined requirements and behaves as expected. This iterative process of writing tests, making them pass, and then refactoring the code for improvement is at the heart of TDD. It fosters a culture of continuous improvement, where code is always in a state that is ready for further development.

Moreover, TDD facilitates the development of isolated, testable units of code, making it easier to identify and fix issues early in the development process. This is particularly beneficial in Angular applications, where components, services, and modules often have complex dependencies and interactions. By testing these units in isolation, developers can ensure that each part of the application works correctly before integrating them into the larger system.

Furthermore, TDD promotes the development of high-quality, maintainable code by encouraging developers to write clear, concise, and well-documented tests. These tests serve as documentation, making it easier for other developers (or even the original developers in the future) to understand the purpose and functionality of the code.

In the context of Angular, TDD can be particularly effective in developing services, components, and pipes, as demonstrated in the examples provided. By starting with a clear definition of what the code should do, developers can write tests that guide the implementation process, ensuring that the code meets the desired specifications. This approach not only leads to better-designed code but also makes the development process more efficient and enjoyable.

In summary, refactoring and improving Angular code through TDD is a valuable practice that can significantly enhance the quality of your Angular applications. By adopting TDD, developers can ensure that their code is robust, maintainable, and ready for future enhancements. This approach not only benefits the current development cycle but also sets a solid foundation for future development efforts, making it a worthwhile investment for any Angular developer.

This is the end of the book. This book serves as a comprehensive guide for developers looking to adopt or improve their TDD practices in Angular projects. By following the principles and techniques described in this book, developers can significantly improve the reliability, performance, and maintainability of their Angular applications. The book emphasizes the value of a thorough testing strategy, the importance of using the right tools and practices, and the benefits of adopting a test-first approach to development. Whether you're new to Angular or an experienced developer looking to hone your skills, this book offers valuable insights and practical advice for mastering TDD in Angular.

Index

packtpub.com

Subscribe to our online digital library for full access to over 7,000 books and videos, as well as industry leading tools to help you plan your personal development and advance your career. For more information, please visit our website.

Why subscribe?

- Spend less time learning and more time coding with practical eBooks and Videos from over 4,000 industry professionals

- Improve your learning with Skill Plans built especially for you

- Get a free eBook or video every month

- Fully searchable for easy access to vital information

- Copy and paste, print, and bookmark content

Did you know that Packt offers eBook versions of every book published, with PDF and ePub files available? You can upgrade to the eBook version at packtpub.com and as a print book customer, you are entitled to a discount on the eBook copy. Get in touch with us at customercare@packtpub.com for more details.

At www.packtpub.com, you can also read a collection of free technical articles, sign up for a range of free newsletters, and receive exclusive discounts and offers on Packt books and eBooks.

Other Books You May Enjoy

If you enjoyed this book, you may be interested in these other books by Packt:

Angular Design Patterns and Best Practices

Alvaro Camillo Neto

ISBN: 978-1-83763-197-1

- Discover effective strategies for organizing your Angular project for enhanced efficiency.
- Harness the power of TypeScript to boost productivity and the overall quality of your Angular project.
- Implement proven design patterns to streamline the structure and communication between components.
- Simplify complex applications by integrating micro frontend and standalone components.
- Optimize the deployment process for top-notch application performance.
- Leverage Angular signals and standalone components to create performant applications.

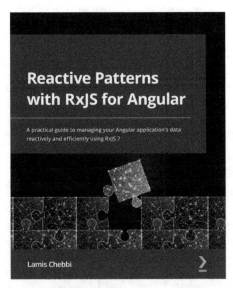

Reactive Patterns with RxJS and Angular Signals

Lamis Chebbi

ISBN: 978-1-83508-770-1

- Get to grips with RxJS core concepts such as Observables, subjects, and operators.
- Use the marble diagram in reactive patterns.
- Delve into stream manipulation, including transforming and combining them.
- Understand memory leak problems using RxJS and best practices to avoid them.
- Build reactive patterns using Angular Signals and RxJS.
- Explore different testing strategies for RxJS apps.
- Discover multicasting in RxJS and how it can resolve complex problems.
- Build a complete Angular app reactively using the latest features of RxJS and Angular.

Packt is searching for authors like you

If you're interested in becoming an author for Packt, please visit `authors.packtpub.com` and apply today. We have worked with thousands of developers and tech professionals, just like you, to help them share their insight with the global tech community. You can make a general application, apply for a specific hot topic that we are recruiting an author for, or submit your own idea.

Hi,

I am Ezéchiel Amen AGBLA author of *Mastering Angular Test-Driven Development*. I really hope you enjoyed reading this book and found it useful for increasing your productivity and efficiency.

It would really help me (and other potential readers!) if you could leave a review on Amazon sharing your thoughts on this book.

Go to the link below or scan the QR code to leave your review:

`https://packt.link/r/1805126083`

Your review will help me to understand what's worked well in this book, and what could be improved upon for future editions, so it really is appreciated.

Best Wishes,
Ezéchiel Amen AGBLA

Download a free PDF copy of this book

Thanks for purchasing this book!

Do you like to read on the go but are unable to carry your print books everywhere?

Is your eBook purchase not compatible with the device of your choice?

Don't worry, now with every Packt book you get a DRM-free PDF version of that book at no cost.

Read anywhere, any place, on any device. Search, copy, and paste code from your favorite technical books directly into your application.

The perks don't stop there, you can get exclusive access to discounts, newsletters, and great free content in your inbox daily

Follow these simple steps to get the benefits:

1. Scan the QR code or visit the link below

https://packt.link/free-ebook/9781805126089

2. Submit your proof of purchase
3. That's it! We'll send your free PDF and other benefits to your email directly

www.ingramcontent.com/pod-product-compliance
Lightning Source LLC
LaVergne TN
LVHW081522050326
832903LV00025B/1585